# Guide to Concrete Repair

*W. Glenn Smoak*

United States Department of the Interior
Bureau of Reclamation
Technical Service Center

Reprint 1998
Reprint 2001
Reprint 2005

April 1997

This guide contains the expertise of numerous individuals who have directly assisted the author on many concrete repair projects or freely shared their concrete repair knowledge whenever requested. Their substantial contributions to the preparation of this guide are acknowledged and appreciated. Some of the material in this guide originated in the various editions of Reclamation's *Concrete Manual*. The author edited, revised, or updated this information for inclusion herein.

Individuals who have been especially helpful to the author include James E. Backstrom, former Reclamation engineer, mentor, and friend, deceased; Edward M. Harboe, Reclamation engineer, retired; U. Marlin Cash, Reclamation technician, deceased; Dennis O. Arney, Reclamation technician, retired; G.W. DePuy, Reclamation engineer, former supervisor and friend, retired; and Kurt D. Mitchell, Reclamation technician. Dr. Dave Harris, Manager, Materials Engineering and Research Laboratory, obtained much of the funding to prepare this guide; Kurt F. Von Fay, Civil Engineer, Materials Engineering and Research Laboratories, performed the peer review; James E. McDonald, Structures Laboratory, Waterways Experiment Station, U.S. Army Corps of Engineers, provided editorial reviews of selected information and many useful suggestions and participated with the author in several cooperative Reclamation—U.S. Corps of Engineers concrete repair programs. The assistance of these and numerous other engineers and technicians is gratefully acknowledged.

# Contents

# Figures

# Chapter I

# Repair of Concrete

**1. Introduction.**—For many years, the Bureau of Reclamation (Reclamation) has published the *Concrete Manual*, the first edition dated July 1938, and more recently, the *Standard Specifications For Repair of Concrete, M-47*, the first edition dated November 1970. The subsequent revisions of these two documents (Bureau of Reclamation, 1975 and 1996), particularly chapter 7 of the *Concrete Manual*, have formed the basis for nearly all concrete repair performed on Reclamation projects during the past 25 years.

Reclamation operates and maintains a water resources infrastructure, located primarily in the harsh climatic zones of the Western United States, valued at over $17 billion. It has become apparent that there is need for modernization and expansion of the information on the methods, materials, and procedures of concrete repair originally found in chapter 7 of the *Concrete Manual*. This *Guide to Concrete Repair* results from recognition of that need. It is designed to serve as a companion document to the "Standard Specifications for Repair of Concrete" included in appendix A of this guide.

This guide first discusses Reclamation's methodology for concrete repair. It then addresses the more common causes of damage to Reclamation concrete, including suggestions of the types of repair methods and materials most likely to be successful in repairing concrete damage resulting from those causes. Finally, the guide contains a detailed description of the uses, limitations, materials, and procedures of each of the standard repair methods/materials included in the "Standard Specifications for Repair of Concrete."

**2. Maintenance of Concrete.**—Modern concrete is a very durable construction material and, if properly proportioned and placed, will give very long service under normal conditions. Many Reclamation concrete structures, however, were constructed using early concrete technology, and they have already provided well over 50 years of service under harsh conditions. Such concrete must be inspected regularly to ensure that it is receiving the maintenance necessary to retain serviceability. Managers and foremen of operation and maintenance crews must understand that, with respect to concrete, there is no such thing as economical deferred maintenance. Failure to promptly provide the proper necessary maintenance will simply result in very expensive repairs or replacement of otherwise useful structures. Figures 1 and 2 demonstrate the folly of inadequate or inappropriate maintenance. These two structures now require replacement at a cost tens of times greater than that of the preventive maintenance that could have extended their serviceability indefinitely.

Experience has shown that there are certain portions of exposed concrete structures more vulnerable than others to deterioration from weathering in freezing climates. These are exposed surfaces of the top 2 feet of walls, piers, posts, handrails, and parapets; all of curbs, sills, ledges, copings, cornices, and corners; and surfaces in contact with spray or water at frequently changing levels during freezing weather. The durability of these surfaces can be considerably improved and serviceability greatly prolonged by preventive maintenance such as weatherproofing treatment with concrete sealing compounds (sections 35 and 38).

Figure 1.—Lack of maintenance has resulted in near loss of this irrigation structure.

Figure 2.—Deferred maintenance has allowed freezing and thawing deterioration
to seriously damage this structure.

Selecting the most satisfactory protective treatment depends to a considerable extent upon correctly assessing the exposure environment. Concrete sealing compounds and coatings that provide good protection from weathering in an essentially dry environment may perform poorly in the presence of an abundance of water such as on some bridge curbs and railings, stilling basin walls, and piers. Freezing and thawing tests of concrete specimens protected by a variety of concrete sealing compounds and coatings, including linseed oil, fluosilicates, epoxy and latex paints, chlorinated rubber, and water-proofing and penetrating sealers, have been performed in Reclamation laboratories. These tests indicate that proprietary epoxy formula-tions, siloxane and silane formulations, and the high molecular weight methacrylate formulations (section 35) clearly excel in resisting deterioration caused by repeated freezing and thawing in the presence of water. None of these formulations, however, will totally "waterproof" concrete. That is, they will not prevent treated concrete from absorbing water and becoming saturated under conditions of complete and long-term submergence.

The performance of new concrete sealing compounds is continually being evaluated by the Materials Engineering and Research Laboratory, Code D-8180, located in Denver, Colorado. If use of these materials is being considered, the project should contact the Denver Office for the latest recommendations on materials, methods of mixing, application, curing, and precautions to be exercised during placement.

Except for hand-placed mortar restorations of deteriorated concrete (section 25), concrete sealing compounds are ordinarily not applied on new concrete construction. The treatments are most commonly used on older surfaces when the earliest visible evidence of weather-ing appears. That is, the treatment is best used before deterioration advances to a stage where it cannot be arrested. Such early evidence consists primarily of fine surface cracking, close and parallel to edges and corners. The need for protection also may be indicated by pattern cracking, surface scaling or spalling, and shrinkage cracking. By treatment of these vulnerable surfaces in the early stages of deterioration, later repairs may be avoided or at least postponed for a long time.

Linseed oil-turpentine-paint preparations have been widely used in the past by Reclamation to retard concrete deterioration caused by weathering. These preparations, when applied correctly, have been effective. The terminology "linseed oil treatment," however, has caused many users to believe that a simple coating of boiled linseed oil would protect concrete from weathering. Such is not the case. The treatment recommended by Reclamation consisted of a number of steps including acid washing surface preparation, 48-hour drying, and application of two or more coats of a hot linseed oil-turpentine mixture followed by two or more coats of white lead paint, the first of which was thinned with linseed oil and turpentine. The modern concrete sealing compounds are much simpler to apply and provide superior protection to the concrete. The use of the linseed oil-turpentine-oil paint system is no longer recommended.

**3. General Requirements for Quality Repair.**—The term "concrete repair" refers to any replacing, restoring, or renewing of concrete or concrete surfaces after initial placement. The need for repairs can vary from such minor imperfections as she-bolt holes, snap-tie holes, or normal weathering to major damages resulting from water energy or structural failure. Although the procedures described may initially appear to be unnecessarily detailed, experience has repeatedly demonstrated that no step in a repair operation can be omitted or carelessly performed without detriment to the serviceability of the work. Inadequate workmanship, procedures, or materials will result in inferior repairs which will ultimately fail at significant cost.

*(a) Workmanship.*—It is the obligation of the construction contractor or operation and

maintenance crew to repair imperfections or damage in concrete so that repairs will be serviceable and of a quality and durability comparable to the adjacent portions of the structure. Repair personnel are responsible for making repairs that are inconspicuous, durable, and well bonded to existing surfaces. Since most repair procedures involve predominantly manual operations, it is particularly important that both foremen and workmen be fully instructed concerning procedural details of repairing concrete and the reasons for the procedures. Workmen should also be apprised of the more critical aspects of repairing concrete. Constant vigilance must be exercised by the contractor's and/or the Government's forces to ensure maintenance of the necessary standards of workmanship. Employment of dependable and capable workmen is essential. Well-trained, competent workmen are particularly essential when epoxy, polyurethane, or other resinous materials are used in repair of concrete.

*(b) Procedures.*—Serviceable concrete repairs can result only if correct methods are chosen and techniques are carefully performed. Wrong or ineffective repair or construction procedures, coupled with poor workmanship, lead to inferior repairs. Many proven procedures for making high quality repairs are detailed in this guide; however, not all procedures used in repair or maintenance are discussed. Therefore, it is incumbent upon the craftsmen doing the work to use procedures that have been successful or that have a proven high reliability factor.

Repairs made on new or old concrete should be made as soon as possible after such need is realized and evaluated. On new work, the repairs that will develop the best bond and, thus, are the most likely to be as durable and permanent as the original work are those that are made immediately after stripping of the forms while the concrete is quite green. For this reason, repairs to newly constructed concrete should be completed within 24 hours after the forms have been removed.

Before repairs are commenced, the method and materials proposed for use should be approved by an authorized inspector. Routine curing should be interrupted only in the area of repair operations.

Effective repair of deteriorated portions of concrete structures cannot be ensured unless there is complete removal of all deteriorated or possibly affected concrete, careful replacement in strict accordance with a standard or approved procedure, and assurance of secure anchorage and effective drainage when needed. Consequently, work of this type should not be undertaken unless or until ample time, personnel, and facilities are available. Only as much of this work should be undertaken as can be completed correctly; otherwise, the work should be postponed, but not so long as to allow further deterioration. Repairs should be made at the earliest possible date.

*(c) Materials.*—Materials to be used in concrete repair must be high quality, relatively fresh, and capable of meeting specifications requirements for the particular application or intended use. Mill reports or testing laboratory reports should be required of the supplier or manufacturer as an indication of quality and suitability. Short of this requirement, certifications stating that the materials meet certain specifications should be required of the supplier. Due to the high cost associated with the subsequent removal and replacement of new, unknown, or unproven materials if they prove unsuitable for the job, such materials should never be used in concrete repair unless (1) the standard repair materials have been determined unsuitable and (2) the owners and all other parties to the repair have been informed of the need to use nonstandard materials and the associated risk.

Materials selected for repair application must be used in accordance with manufacturers' recommendations or other approved methods. Mixing, proportioning, and handling must be in accordance with the highest standards of workmanship.

# Chapter II

# A Concrete Repair System

Concrete repairs have occurred on Reclamation projects since the first construction concrete was placed in 1903. Unfortunately, even though the best available materials were used, many repair failures have occurred during the 90 years since that first concrete construction. In evaluating the causes of these failures, it was learned that it is essential to consistently use a systematic approach to concrete repair. There are several such repair approaches or systems currently in use. The U.S. Army Corps of Engineers lists an excellent system in the first chapter of its manual, *Evaluation and Repair of Concrete Structures* (U.S. Army Corps of Engineers, 1995). Other organizations, such as the American Concrete Institute, the Portland Cement Association, the International Concrete Repair Institute, and private authors (Emmons, 1994) have also published excellent methodologies for concrete repair. This guide will not attempt to discuss or evaluate these systems for any particular set of field conditions. Rather, the following seven-step repair system, which has been developed, used, and evaluated by Reclamation over an extended period of time, is presented. This methodology has been found suitable for repairing construction defects in newly constructed concrete as well as old concrete that has been damaged by long exposure and service under field conditions.

This system will be found most useful if followed in a numerically sequential, or step wise manner. Quite often, the first questions asked when the existence of deteriorated or damaged concrete becomes apparent are: "What should be used to repair this?" and "How much is this going to cost?". These are not improper questions. However, they are questions asked at an improper time. Ultimately, these questions must be answered, but pursuing answers to these questions too early in the repair process will lead to incorrect and, therefore, extremely costly solutions. If a systematic approach to repair is used, such questions will be asked when sufficient information has been developed to provide correct and economical answers.

*Reclamation's Concrete Repair System*

1. Determine the cause(s) of damage
2. Evaluate the extent of damage
3. Evaluate the need to repair
4. Select the repair method
5. Prepare the old concrete for repair
6. Apply the repair method
7. Cure the repair properly

**4. Determine the Cause(s) of Damage.—** The first and often most important step of repairing damaged or deteriorated concrete is to correctly determine the cause of the damage. If the cause of the original damage to concrete is not determined and eliminated, or if an incorrect determination is made, whatever damaged the original concrete will likely also damage the repaired concrete. Money and effort spent for such repairs is, thus, totally wasted. Additionally, larger and even more costly replacement repairs will then be required.

If the original damage is the result of a one-time event, such as a river barge hitting a bridge pier, an earthquake, or structural overload, remediation of the cause of damage need not be addressed. It is unlikely that such an event will occur again. If, however, the cause of damage is of a continuing or recurring nature, remediation must be addressed, or the repair method and materials must in some manner be made resistant to predictable future damage. The more common causes of damage to Reclamation concrete are discussed in chapter III. A quick review of these common causes of damage reveals that the majority of them are of a continuing or recurring nature.

It is important to differentiate between causes of damage and symptoms of damage. In the above case of the river barge hitting the bridge pier, the cause of damage is the impact to the concrete. The resultant cracking is a symptom of that impact. In the event of freezing and thawing deterioration to modern concrete, the cause of the damage may well lie with the use of low quality or dirty fine or coarse aggregate in the concrete mix. The resultant scaling and cracking is a symptom of low durability concrete. The application of high cost repairs to low quality concrete is usually economically questionable.

It is somewhat common to find that multiple causes of damage exist (section 23). Improper design, low quality materials, or poor construction practices reduce the durability of concrete and increase its susceptibility to deterioration from other causes. Similarly, sulfate and alkali-silica deterioration cause cracks in the exterior surfaces of concrete that allow accelerated deterioration from cycles of freezing and thawing. The deterioration resulting from the lowered resistance to cyclic freezing and thawing might mask the original cause of the damage.

Finally, it is important to fully understand the original design intent and concepts of a damaged structure before attempting repair. Low quality local aggregate may have intentionally been used in the concrete mix because the costs associated with hauling higher quality aggregate great distances may have made it more economical to repair the structure when required at some future date. A classic example of misunderstanding the intent of design recently occurred on a project in Nebraska. A concrete sluiceway that would experience great quantities of waterborne sand was designed with an abrasion-resistant protective overlay of silica fume concrete. This overlay was intentionally designed so that it would not bond to the base concrete, making replacement easier when required by the anticipated abrasion-erosion damage. This

design concept, however, was not communicated to construction personnel who became deeply concerned when the silica fume overlay was found to be "disbonded" shortly after placement and curing was completed. Some difficulty was experienced in preventing field personnel from requiring the construction contractor to repair a perfectly serviceable overlay that was performing exactly as intended.

**5. Evaluate the Extent of Damage.**—The next step of the repair process is to evaluate the extent and severity of damage. The intent of this step is to determine how much concrete has been damaged and how this damage will affect serviceability of the structure (how long, how wide, how deep, and how much of the structure is involved). This activity includes prediction of how quickly the damage is occurring and what progression of the damage is likely.

The importance of determining the severity of the damage should be understood. Damage resulting from cyclic freezing and thawing, sulfate exposure, and alkali-aggregate reaction appears quite similar. The damage caused by alkali-aggregate reaction and sulfates is far more severe than that caused by freeze-thaw, although all three of these causes can result in destruction of the concrete and loss of the affected structure. The main difference in severity lies in the fact that proper maintenance can reduce or eliminate damage caused by freeze-thaw. There is no proven method of reducing damage caused by alkali-aggregate reaction or sulfate exposure.

The most common technique used to determine the extent of damage is sounding the damaged and surrounding undamaged concrete with a hammer. If performed by experienced personnel, this simple technique, when combined with a close visual inspection, will provide the needed information in many instances of concrete damage. In sounding suspected delaminated or disbonded concrete, it should be remembered that deep delaminations or delaminations that contain only minute separation may not always sound drummy or hollow. The presence of such delaminations can be detected by placing a hand close to the location of hammer blows or by closely observing sand particles on the surface close to the hammer blows. If the hand feels vibration in the concrete, or if the sand particles are seen to bounce however slightly due to the hammer blows, the concrete is delaminated.

An indication of the strength of concrete can also be determined by hammer blows. High strength concrete develops a distinct ring from a hammer blow and the hammer rebounds smartly. Low strength concrete resounds with a dull thud and little rebound of the hammer. More detailed information can be obtained by using commercially available rebound hammers, such as the Schmidt Rebound Hammer.

Cores taken from the damaged areas can be used to detect subsurface deterioration, to determine strength properties through laboratory testing, and to determine petrography. Petrographic examination of concrete obtained by coring can also be very useful in determining some causes of deterioration.

There are a number of nondestructive testing methods that can be used to evaluate the extent of damage (Poston et al., 1995). The above-mentioned Schmidt Rebound Hammer is perhaps the cheapest and simplest to use. Ultrasonic pulse velocity and acoustic pulse echo devices measure the time required for an electronically generated sound wave to either travel through a concrete section or to travel to the far side of a concrete section and rebound. Damaged or low quality concrete deflects or attenuates such sound waves and can be detected by comparison of the resulting travel time with that of sound concrete. Acoustic emission devices detect the elastic waves that are generated when materials are stressed or strained beyond their elastic limits. With such devices, it is possible to "hear" the impulses from development of microcracks in overly stressed concrete. Acoustic emission equipment has been used to "hear" the occurrence of prestressing strand failure in large diameter prestressed concrete pipe.

With computer assistance, several acoustic emission devices have been used not only to detect the occurrence of strand failure(s), but through triangulation, they were able to determine the location of the failure(s) (Travers, 1994).

The areas of deteriorated or damaged concrete discovered by these methods should be mapped or marked on drawings of the affected structure to provide information needed in subsequent calculations of the area and volume of concrete to be repaired and for preparation of repair specifications. Even though care is taken in these investigations, it is common to find during preparation of the concrete for repair that the actual area and volume of deteriorated concrete exceeds the original estimate. For this reason, it is usually a good idea to increase the computed quantity estimates by 15 to 25 percent to cover anticipated overruns.

**6. Evaluate the Need to Repair.**—Not all damaged concrete requires immediate repair. Many factors need consideration before the decision to perform repairs can be made. Obviously, repair is required if the damage affects the safety or safe operation of the structure. Similarly, repairs should be performed if the deterioration has reached a state, or is progressing at a rate, such that future serviceability of the structure will be reduced. Most concrete damage, however, progresses slowly, and several options are usually available if the deterioration is detected early. With early detection, it may be possible to arrest the rate of deterioration using maintenance procedures. Even if repair is required, early detection of damage will allow orderly budgeting of funds to pay the costs of repair.

Some types of concrete deterioration can simply be ignored. Cracking due to drying shrinkage and freezing and thawing deterioration is common on the downstream face of many older western dams. These types of damage are unsightly, but repair can seldom be justified for other than cosmetic purposes. It should be anticipated that such repairs might be more unsightly and of lower durability than the existing concrete. Conversely, structural cracks due to foundation settlement and freezing and thawing deterioration to the walls or floor of a spillway will usually require repair, if not immediately, at some point in the future. Figure 3 shows freezing and thawing damage to the face of a dam that does not require repair for safe operation of the structure. Figure 4 shows similar damage that should have been repaired long ago. Damage caused by absorptive aggregate popouts is common on bridge deck, canal, and dam concrete (figure 5). Unless such concrete is exposed to high velocity waterflows, where the offsets caused by popouts can result in cavitation damage, repair can be ignored. Figure 6 shows damage to a spillway that appears quite serious, and repair is obviously required. This spillway, however, is constructed with a very thick slab and does not experience high velocity water flow. The repairs can be scheduled at some future date to allow an orderly process of budgeting to obtain the required funding. It should be noted, however, that proper maintenance might have eliminated the need to repair this spillway.

Selecting or scheduling the most optimum time to perform needed concrete repair should be part of the process of determining the need to repair. Except in emergencies, many irrigation structures cannot be removed from service during the water delivery season. The expense or loss of income involved with the inopportune release of reservoir water in order to lower water surface elevations to accomplish repairs may exceed the costs of the repairs by many times. If such costs exceed the value of the benefits expected from performing repairs, it might be prudent to postpone or even cancel performance of the repairs. Figure 7 shows damage on a spillway floor. This damage was initially judged to be of a nonserious nature. Closer evaluation, however, revealed that foundation material had been removed from a very large area beneath this floor slab and that immediate repair was required. Had this spillway been operated without repair during periods of high spring runoff or floodflows, extensive additional damage might have resulted.

These first steps—determining the cause of damage, evaluating the extent of damage, and evaluating the need to repair—form the basis of what is known as a condition survey. If the damage is not extensive or if only a small part of a structure is involved, the condition survey could be simply a mental exercise. If major repair or rehabilitation is required, a detailed condition survey should be performed and documented. Such a survey will consist of review of the plans, specifications, and operating parameters for the structure; determination of concrete properties; and any additional field surveys, engineering studies, or structural analysis required to fully evaluate the present and desired conditions of the structure (American Concrete Institute, 1993). The final feature of a condition survey, completed only after the above-listed items have been completed, is a list of the recommended repair methods and materials.

**7. Select the Repair Method.**—There is a tendency to attempt selection of repair methods/materials too early in the repair process. This should be guarded against. With insufficient information, it is very difficult to make proper, economical, and successful selections. Once the above three steps of the repair process have been completed, or upon completion of a detailed condition survey, the selection of proper repair methods and materials usually becomes very easy. These steps define the types of conditions the repair must resist, the available repair construction time period, and when repairs must be accomplished. This information, in combination with data on the volume and area of concrete to be repaired, will usually determine which of the 15 standard repair materials should be used. Also, this information will determine when the standard repair materials cannot be expected to perform well and when nonstandard materials should be considered (see chapter V). Chapter IV contains a detailed discussion of each of the standard repair materials.

**8. Prepare the Old Concrete for Repair.**— Preparation of the old concrete for application of the repair material is of primary importance in the accomplishment of durable repairs. The

Figure 3.—Freezing and thawing deterioration to the downstream face of this dam does not require repair for safe operation of the structure.

Figure 4.—This freezing and thawing deterioration should have been repaired before it advanced to the point that wall replacement or removal is the only option.

Figure 5.—Absorptive aggregate popout on spillway floor.

Figure 6.—Spillway damage requiring repairs at some future date.

Figure 7.—This concrete damage was found to be a serious threat to the structural integrity of this spillway.

very best of repair materials will give unsatisfactory performance if applied to weakened or deteriorated old concrete. The repair material must able to bond to sound concrete. It is essential that all of the unsound or deteriorated concrete be removed before new repair materials are applied.

*Saw Cut Perimeters.* The first step in preparing the old concrete for repair is to saw cut the perimeter of the repair area to a depth of 1 to 1.5 inches. The purpose of the saw cuts is to provide a retaining boundary against which the repair material can be compacted and consolidated. The perimeters of repairs are the locations most exposed to the effects of shrinkage, deterioration, and bond failure. Only poor compaction of repair material can be accomplished at feather edge perimeters. Such repair zones will fail quickly. For this

reason feather edge perimeters to repair areas are not permitted by Reclamation's M-47 specifications. It is unnecessary to cut to the full depth of the repair, although to do so is not harmful. The saw cuts should be perpendicular to the concrete surface or tilted inward 2 to 3 degrees to provide retaining keyways that mechanically lock the repair material into the area. Tilting the saw inward more than 3 degrees may result in weak top corners in the old concrete and should be avoided. The saw cuts should never be beveled outward.

It is usually false economy to try to closely follow the shape of the repair area with a multitude of short saw cuts as seen at the bottom of figure 8. The cost of sawing such a shape most likely will exceed the cost of increased repair area, and the resulting repair

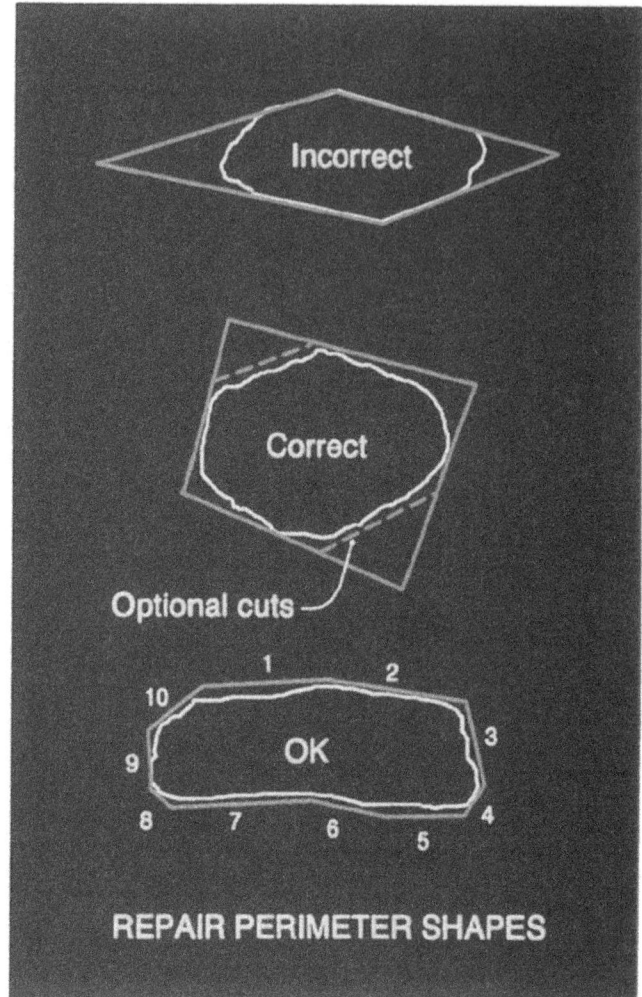

Figure 8.—Saw cut patterns for the perimeters of repair areas.

may be less attractive than those having simple rectangular shapes. Saw cuts should not meet in acute angles as shown at the top of figure 8. It is very difficult to compact repair material into such sharp corners. The saw cut perimeters should have rounded corners, as seen in figure 9, whenever reasonable. Rounded corners cannot be cut with a circular concrete saw, but the cuts can be stopped short of the intersection and rounded using a jackhammer or bush hammer carefully held in a vertical orientation. It should be noted that intersections cannot be cut with a circular saw without the cuts extending outside the intersection. These cut extensions often serve as sources of cracking in some repair materials. Once the perimeters have been cut, the deteriorated concrete is removed using methods discussed in following paragraphs.

*Concrete Removal.* All deteriorated or damaged concrete must be removed from the repair area to provide sound concrete for the repair material to bond to. It is always false economy to attempt to save time or money by shortchanging the removal of deteriorated concrete. Whenever possible, the first choice of concrete removal technique should be high pressure (8,000 to 15,000 pounds per square inch [psi]) hydroblasting or hydrodemolition. These techniques have the advantage of removing the unsound concrete while leaving high quality concrete in place. They have a further advantage in that they do not leave microfractured surfaces on the old concrete. Impact removal techniques, such as bush-hammering, scrabbling, or jackhammering, can leave surfaces containing a multitude of microfractures which seriously reduce the bond of the repair material to the existing

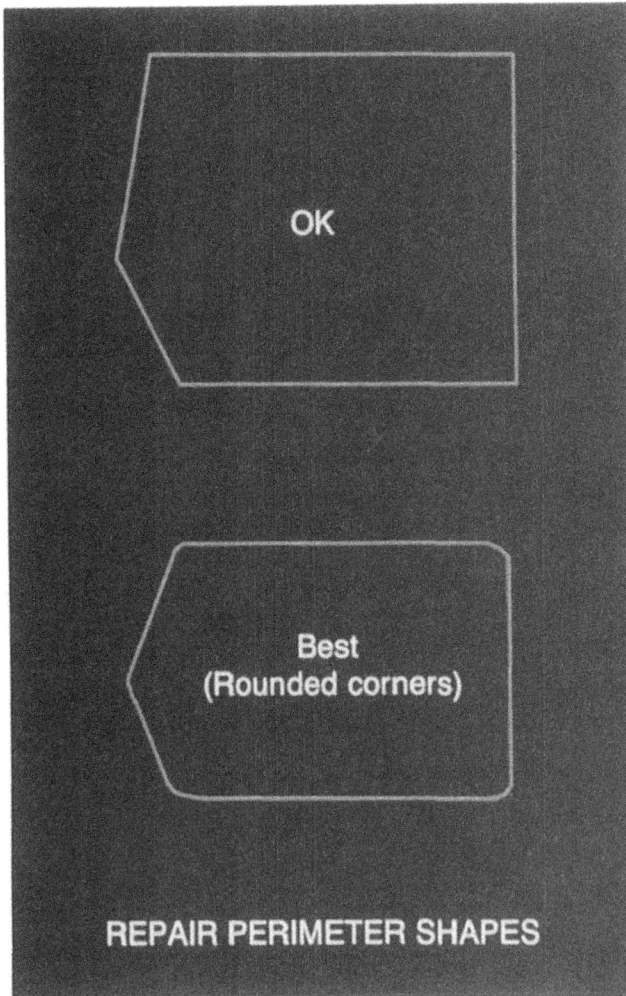

OK

Best
(Rounded corners)

REPAIR PERIMETER SHAPES

Figure 9.—Corners of repair areas
should be rounded whenever possible.

concrete. Subsequent removal of the micro-fractured surface by hydroblasting, shot blasting, or by wet or dry sandblasting is required by Reclamation's M-47 specifications if impact removal techniques are used. A disadvantage of the high pressure water blasting techniques is that the waste water and debris must be handled in an environmentally acceptable manner as prescribed by local regulations.

Impact concrete removal techniques, such as jackhammering for large jobs and bush-hammering for smaller areas, have been used for many years. These removal procedures are quick and economical, but it should be kept in mind that the costs of subsequent removal of the microfractured surfaces resulting from these techniques must be included when comparing the costs of these

techniques to the costs of high pressure water blasting. The maximum size of jackhammers should usually be limited to 60 pounds. The larger jackhammers remove concrete at a high rate but are more likely to damage surrounding sound concrete. The larger hammers can impact and loosen the bond of concrete to reinforcing steel for quite some distance away from the point of impact. Pointed hammer bits, which are more likely to break the concrete cleanly rather than to pulverize it, should be used to reduce the occurrence of surface microfracturing.

Shallow surface deterioration (usually less than 1/2 inch deep) is best removed with shot blasting (figure 10) or dry or wet sand-blasting. Shot blasting equipment is highly efficient and usually includes some type of vacuum pickup of the resulting dust and

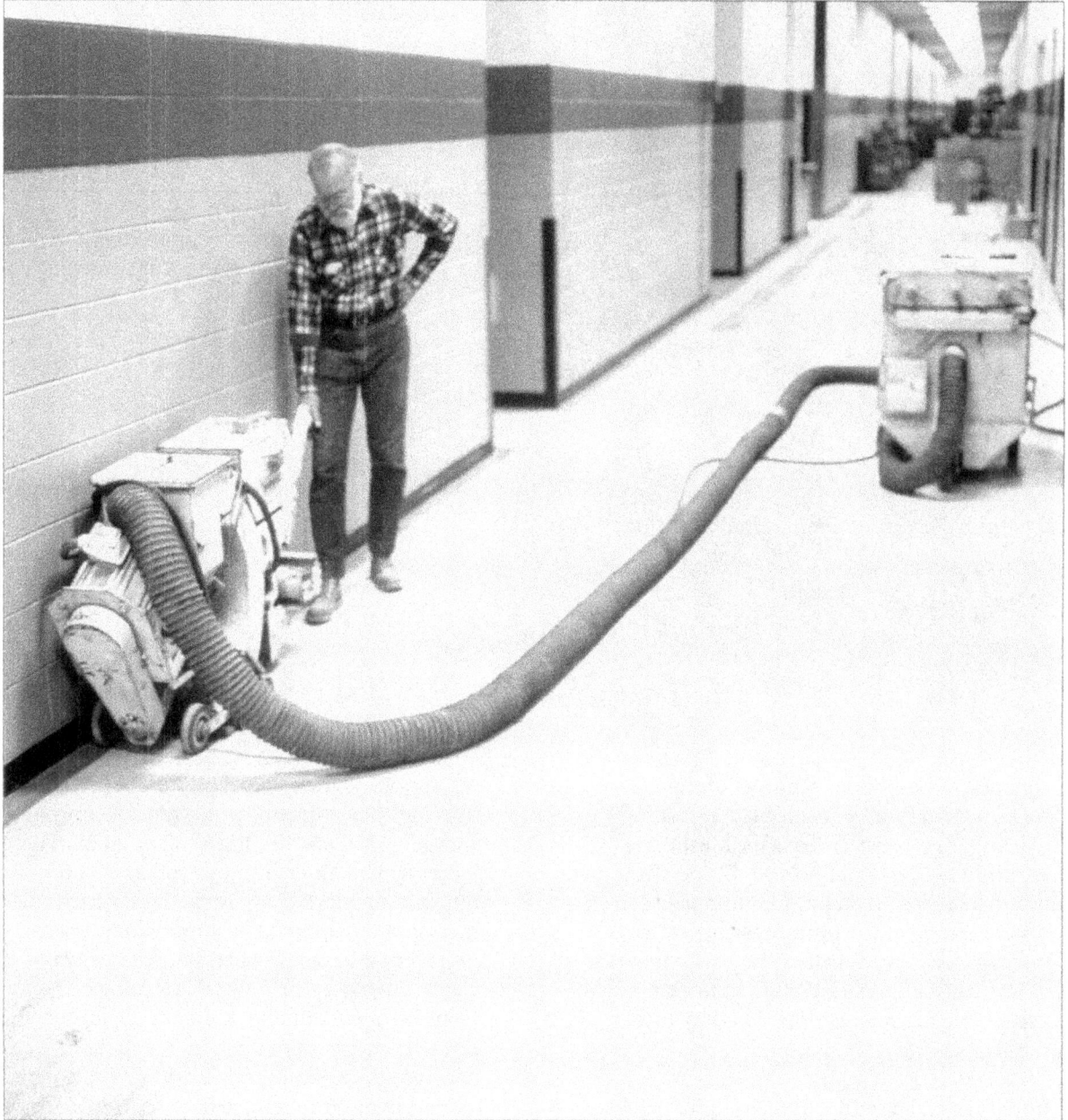

Figure 10.—Shot blasting equipment used to remove shallow concrete deterioration.

debris. The use of such equipment is much more environmentally acceptable than dry sand blasting. The need for removal of such shallow depths of deteriorated concrete is seldom encountered in Reclamation repairs other than for removal of microfractured surfaces or for cosmetic surface cleaning. Shallow deterioration to concrete surfaces can also be removed with tools known as scrabblers (figure 11). These tools usually have multiple bits (figure 12) which pound and pulverize the concrete surfaces in the removal process. Their use greatly multiplies the micro fractures in the remaining concrete surfaces. Extensive high pressure water, sand, or shot blasting efforts are then needed to remove the resulting damaged surfaces. Such efforts are seldom attained under field conditions. For this reason, Reclamation's M-47 specifications prohibit use of scrabblers for concrete removal.

*Reinforcing Steel Preparation.* Reinforcing steel exposed during concrete removal requires special treatment. As a minimum, all scale, rust, corrosion, and bonded concrete must be removed by wire brushing or high pressure water or sand blasting. It is not necessary to clean the steel to white metal condition, just to remove all the loose or poorly bonded debris that would affect bond between the repair material and the reinforcing steel. If corrosion has reduced the cross section of the steel to less than 75 percent of its original diameter, the affected bars should be removed and replaced in accordance with section 12.14 of American Concrete Institute (ACI) 318 (ACI, 1992). Steel exposed more than one-third of its perimeter circumference should be sufficiently exposed to provide a 1-inch minimum clearance between the steel and the concrete. Figure 13 shows the correct concrete removal and preparation for repairing a delamination occurring at the top mat of reinforcing steel of a concrete slab. Figure 14 shows correct preparation of a concrete defect that extends entirely through a wall. Figure 15 shows a properly prepared shallow repair area on a highway bridge deck.

*Maintenance of Prepared Area.* After the repair area has been prepared, it must be maintained in a clean condition and protected from damage until the repair materials can be placed and cured. In hot climates, this might

Figure 11.—Scrabbler equipment used to remove shallow concrete deterioration.

Figure 12.—Multiple bits on the head of a scrabbler pound and pulverize the concrete surface during the removal process.

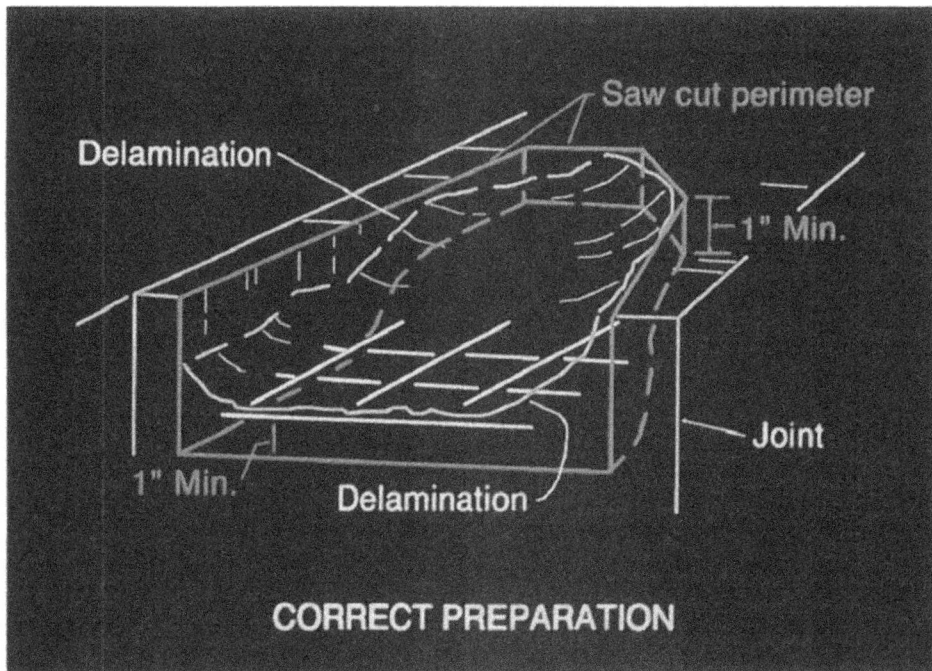

Figure 13.—Correct preparation of a concrete delamination. Perimeter has been saw cut to a minimum depth of 1 inch, and concrete has been removed to at least 1 inch beneath exposed reinforcing steel.

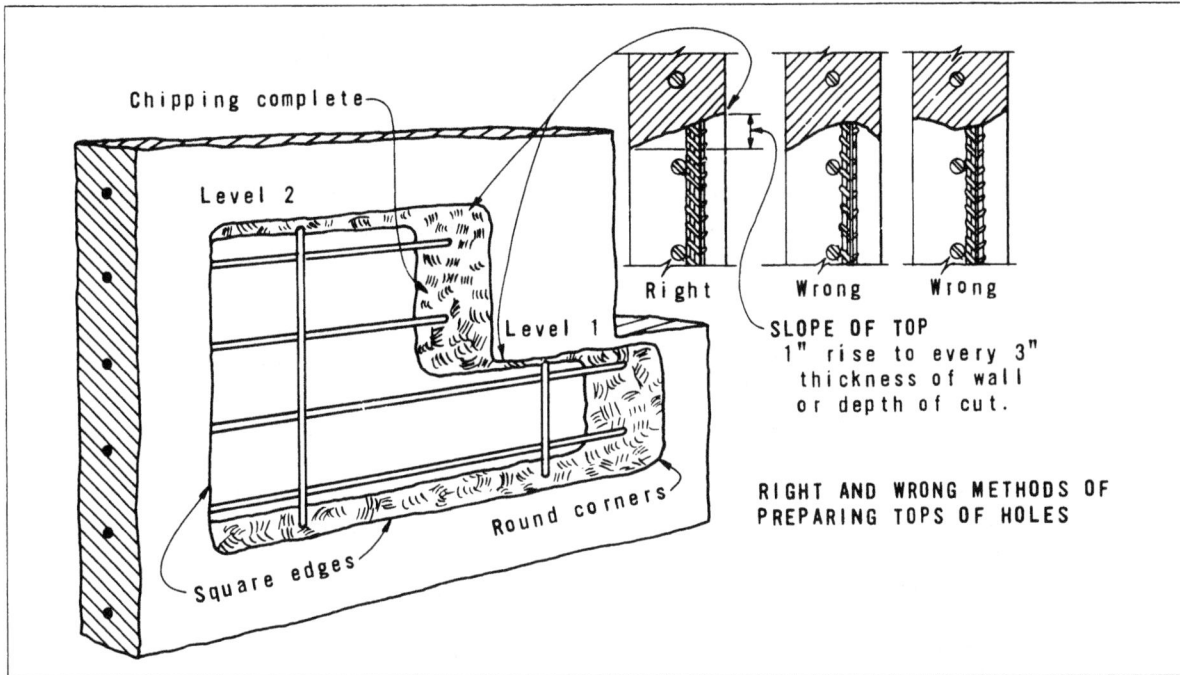

Figure 14.—Preparation of a concrete deterioration that extends completely through a concrete wall.

Figure 15.—Preparation of a shallow defect on a highway bridge deck.

involve providing shade to keep the concrete cool, thereby reducing rapid hydration or hardening. If winter conditions exist, steps need to be taken to provide sufficient insulation and/or heat to prevent the repair area from being covered with snow, ice, or snowmelt water. It should be remembered that repair activities can also contaminate or damage a properly prepared site. Workmen placing repair materials in one area of a repair often track mud, debris, cement dust, or concrete into an adjacent repair area. Once deposited on a prepared surface, this material will serve as a bond breaker if not cleaned up before the new repair material is placed. Repair contractors should be required to repeat preparation if a repair area is allowed to become damaged or contaminated. The prepared concrete should be kept wet or dry, depending upon the repair material to be used. Surfaces that will receive polymer concrete or epoxy-bonded materials should be kept as dry as possible. Some epoxies will bond to wet concrete, but they always bond better to dry concrete. Surfaces that will be repaired with cementitious material should be in a saturated surface dry (SSD) condition immediately prior to material application. This condition is achieved by soaking the surfaces with water for 2 to 24 hours just before repair application. Immediately before material application, the repair surfaces should be blown free of water, using compressed air. The SSD condition prevents the old concrete from absorbing mix water from the repair material and promotes development of adequate bond strength in the repair material. The presence of free water on the repair surfaces during application of the repair material must be avoided whenever practicable.

**9. Apply the Repair Method.**—There are 15 different standard concrete repair methods/materials in Reclamation's M-47 specification. Each of these materials has uniquely different requirements for successful application. These requirements and application procedures are discussed at length in chapter IV of this guide.

**10. Cure the Repair Properly.**—All of the standard repair materials, with the exception of some of the resinous systems, require proper curing procedures. Curing is usually the final step of the repair process, followed only by cleanup and demobilization, and it is somewhat common to find that the curing step has been shortened, performed haphazardly, or eliminated entirely as a result of rushing to leave the job or for the sake of perceived economies. It should be understood that proper curing does not represent unnecessary costs. Rather, it represents a sound investment in long-term insurance. Inadequate or improper curing can result in significant loss of money. At best, improper curing will reduce the service life of the repairs. More likely, inadequate or improper curing will result in the necessity to remove and replace the repairs. The costs of the original repair are, thus, completely lost, and the costs of the replacement repair will be greater because the replacement repairs will be larger and must include the costs of removal of the failed repair material. The curing requirements for each of the 15 standard repair materials are discussed in chapter IV.

# Chapter III

# Causes of Damage to Concrete

The more common causes of damage to Reclamation concrete are discussed in this chapter. The discussion for each cause of damage consists of (1) a description of the cause and how it damages concrete and (2) a discussion and/or listing of appropriate methods/materials to repair that particular type of concrete damage. The format for the text of this chapter was chosen in recognition of the importance of first determining the cause(s) of damage to concrete before trying to select the repair method. It is expected that the full discussion of the selected repair method, as found in chapter IV, will be consulted prior to performance of the work.

**11. Excess Concrete Mix Water.**—The use of excessive water in concrete mixtures is the single most common cause of damage to concrete. Excessive water reduces strength, increases curing and drying shrinkage, increases porosity, increases creep, and reduces the abrasion resistance of concrete. Figure 16 shows the cumulative effects of water-cement ratio on the durability of concrete. In this figure, high durability is associated with low water-cement ratio and the use of entrained air.

Damage caused by excessive mix water can be difficult to correctly diagnose because it is usually masked by damage from other causes. Freezing and thawing cracking, abrasion erosion deterioration, or drying shrinkage cracking, for example, is often blamed for damage to concrete when, in reality, excessive mix water caused the low durability that allowed these other causes to attack the concrete. During petrographic examination, extreme cases of excessive mix water in hardened concrete can sometimes be detected by the presence of bleed water channels or water pockets under large aggregate. More commonly, examination of the batch sheets,

mix records, and field inspection reports will provide confirmation of the use of excessive mix water in damaged concrete. It should be recognized, however, that water added to transit truck mixes at the construction site or applied to concrete surfaces during finishing operations often goes undocumented.

The only permanent repair of concrete damaged by excessive mix water is removal and replacement. However, depending on the extent and nature of damage, a number of maintenance or repair methods can be useful in extending the service life of such concrete. If the damage is detected early and is shallow (less than 1.5 inches deep), application of concrete sealing compounds, such as the high solids content (greater than 15 percent) oligomeric alkyl-alkoxy siloxane or silane systems (section 38) or the high molecular weight methacrylic monomer system (section 35), will reduce water penetration and improve resistance to freeze-thaw spalling and deterioration. Such systems require reapplication at 5- to 10-year intervals. Epoxy-bonded replacement concrete (section 31) can be used to repair damage that extends between 1.5 and 6 inches into the concrete, and replacement concrete (section 29) can be used to repair damage 6 inches deep or deeper.

**12. Faulty Design.**—Design faults can create many types of concrete damage. Discussion of all the types of damage that can result from faulty design is beyond the scope of this guide. However, one type of design fault that is somewhat common is positioning embedded metal such as electrical conduits or outlet boxes too near the exterior surfaces of concrete structures. Cracks form in the concrete over and around such metal features and allow accelerated freeze-thaw deterioration to occur. Bases of handrails or guardrails

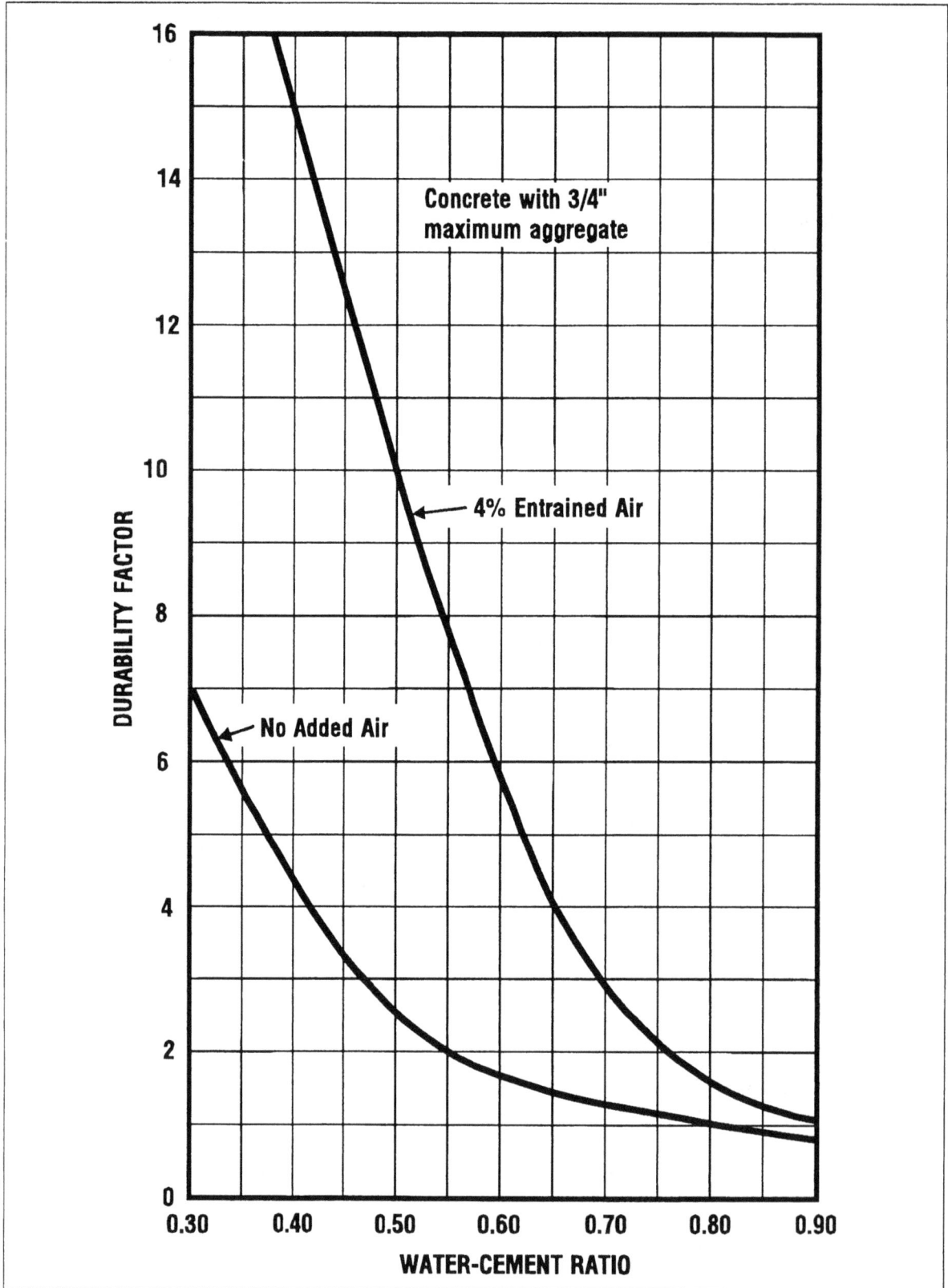

Figure 16.—Relation between durability and water-cement ratio for air entrained and nonair entrained concrete.

are placed too near the exterior corners of walls, walkways, and parapets with similar results. These bases or intrusions into the concrete expand and contract with temperature changes at a rate different from the concrete. Tensile stresses, created in the concrete by expanding metal, cause cracking and subsequent freeze-thaw damage. Long guardrails or handrails can create another problem. The pipe used for such rails also undergoes thermal expansion and contraction. If sufficient slip joints are not provided in the rails, the expansion and contraction cause cracking at the points where the rail attachment bases enter the concrete. This cracking also allows accelerated damage to the concrete from freezing and thawing.

Insufficient concrete cover over reinforcing steel is a common cause of damage to highway bridge structures. This can also be a problem in hydroelectric and irrigation structures. Reclamation usually requires a minimum of 3 inches of concrete cover over reinforcing steel, but in corrosive environments, this can be insufficient. Concrete exposed to the corrosive effects of sulfates, acids, or chlorides should have a minimum of 4 inches of cover to protect the reinforcing steel. Insufficient cover allows corrosion of the reinforcing steel to begin. The iron oxide byproducts of this corrosion require more space in the concrete than the reinforcing steel and result in cracking and delaminating in the concrete.

Failure to provide adequate contraction joints or failure to make expansion joints wide enough to accommodate temperature expansion in concrete slabs will result in damage. Concrete with inadequate contraction joints will crack and make a joint wherever a joint was needed but not provided. Unfortunately, such cracks will not be as visually attractive as a formed or sawed joint. Formation of the cracks relieves the tensile stresses and, though unsightly, seldom requires repair. Concrete slabs constructed with insufficient or too narrow expansion joints can cause serious damage to bridge deck surfaces, dam roadways, and the floors

of long, steeply sloping, south facing spillways. Such concrete experiences large daily and seasonal temperature changes resulting from solar radiation. The resulting concrete expansion is greater in the top surfaces of the slabs, where the concrete temperatures are higher, and less in the cooler bottom edges. Such expansion can cause the upper portions of concrete in adjacent slabs to butt against one another at the joints between the slabs. The only possible direction of relief movement in such slabs is upward, which causes delaminations to form in the concrete, starting at the joints and extending an inch or two back into the slab. These delaminations are commonly located at the top mat of reinforcing steel. In temperate climates, the formation of delaminations relieves the expansion strains, and further damage will usually cease. In cold climates, however, water can enter the delaminations where it undergoes a daily cycle of freezing and thawing. This causes the delaminations to grow and extend as much as 3 to 5 feet away from the joint. Figure 17 is an exaggerated example of such damage.

Repair of damage caused by faulty design is futile until the design faults have been mitigated. Embedded metal features can be removed, handrails can be provided with slip joints, and guardrail attachment bases can be moved to locations with sufficient concrete to withstand the tensile forces. Mitigation of insufficient concrete cover over reinforcing steel is very difficult, but repair materials resistant to those particular types of corrosion can be selected for the repair. Repairs can also be protected by concrete sealing compounds or coatings to reduce water penetration. Slabs containing inadequate expansion joints can be saw cut to increase the number of joints and/or to widen the joints to provide sufficient room for the expected thermal expansion.

Damage caused by design faults can most likely be repaired using replacement concrete (section 29), epoxy-bonded replacement concrete (section 31), or epoxy-bonded epoxy mortar (section 30).

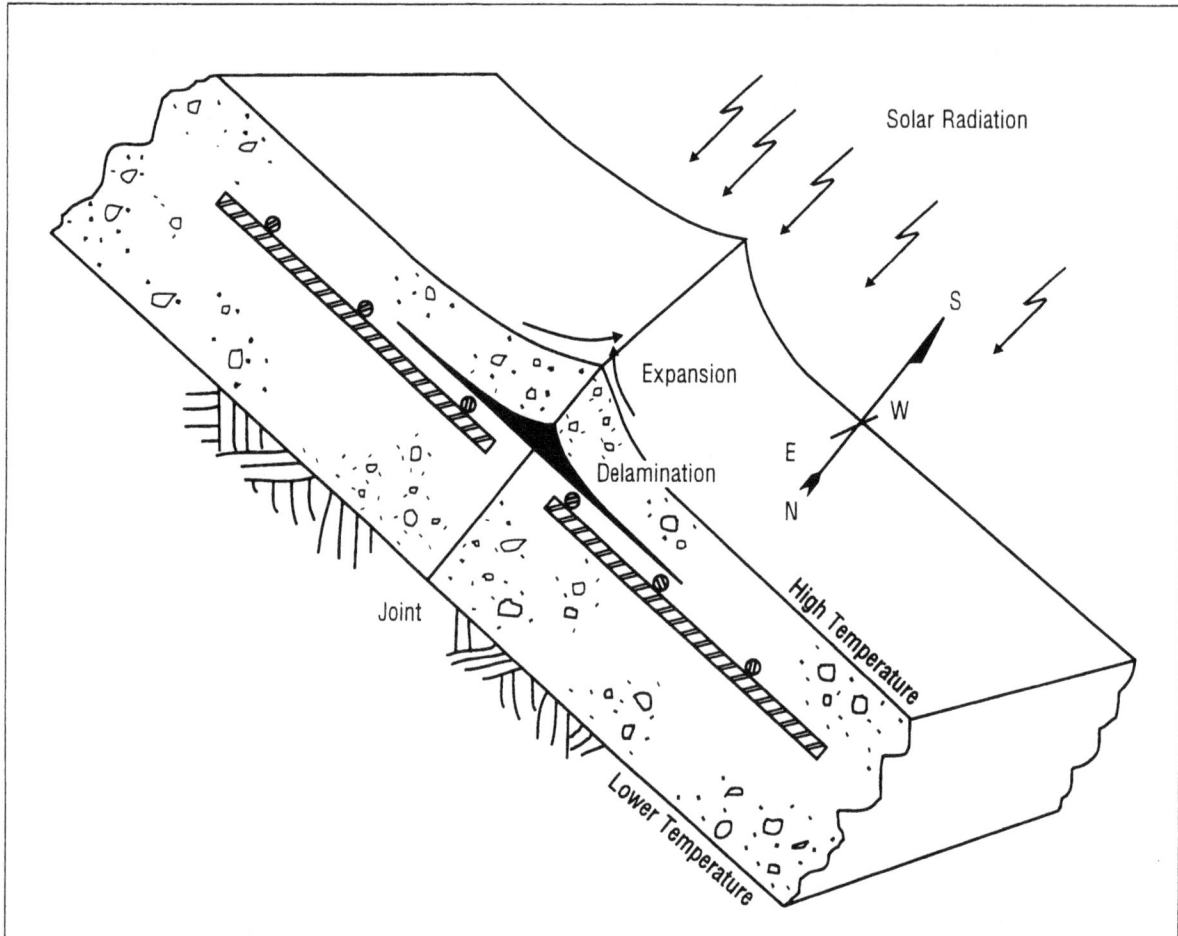

Figure 17.—Delamination caused by solar expansion.

**13. Construction Defects.**—Some of the more common types of damage to concrete caused by construction defects are rock pockets and honeycombing, form failures, dimensional errors, and finishing defects.

Honeycomb and rock pockets are areas of concrete where voids are left due to failure of the cement mortar to fill the spaces around and among coarse aggregate particles. These defects, if minor, can be repaired with cement mortar (section 25) if less than 24 hours has passed since form removal. If repair is delayed longer than 24 hours after form removal, or if the rock pocket is extensive, the area must be prepared and the defective concrete must be removed and replaced with dry pack (section 26), epoxy-bonded replacement concrete (section 31), or replacement concrete (section 29).

Some minor defects resulting from form movement or failure can be repaired with surface grinding (section 24). More likely, the resulting defect is either simply accepted by the owner, or the contractor is required to remove the defective concrete and reconstruct that portion of the structure.

There are many opportunities to create dimensional errors in concrete construction. Whenever possible, it usually is best to accept the resulting deficiency rather than attempt to repair it. If the nature of the deficiency is such that it cannot be accepted, then complete removal and reconstruction is probably the best course of action. Occasionally, dimensional errors can be corrected by removing the defective concrete and replacing it with epoxy-bonded concrete or replacement concrete.

Finishing defects usually involve overfinishing or the addition of water and/or cement to the surface during the finishing procedures. In each instance, the resulting surface is porous and permeable and has low durability. Poorly finished surfaces exhibit surface spalling early in their service life. Repair of surface spalling involves removal of the weakened concrete and replacement with epoxy-bonded concrete (section 31). If the deterioration is detected early, the service life of the surface can be extended through the use of concrete sealing compounds (sections 35 and 38).

**14. Sulfate Deterioration.**—Sodium, magnesium, and calcium sulfates are salts commonly found in the alkali soils and groundwaters of the Western United States. These sulfates react chemically with the hydrated lime and hydrated aluminate in cement paste and form calcium sulfate and calcium sulfoaluminate. The volume of these reaction byproducts is greater than the volume of the cement paste from which they are formed, causing disruption of the concrete from expansion. Type V portland cement, which has a low calcium aluminate content, is highly resistant to sulfate reaction and attack and should be specified when it is recognized that concrete must be exposed to soil and groundwater sulfates. See table 2 of the *Concrete Manual* (Bureau of Reclamation, 1975) for guidance on materials and mixture proportions for concretes exposed to sulfate environments.

Concrete that is undergoing active deterioration and damage due to sulfate exposure can sometimes be helped by application of a thin polymer concrete overlay (section 33) or concrete sealing compounds (sections 35 and 38). Alternate wetting and drying cycles accelerate sulfate deterioration, and some slowing of the rate of deterioration can be accomplished by interrupting the cyclic wetting and drying. Procedures for eliminating or removing waterborne sulfates are also helpful if this is the source of the sulfates. Otherwise, the deteriorating concrete should be monitored for removal and replacement with concrete constructed of type V cement, when appropriate.

**15. Alkali-Aggregate Reaction.**—Certain types of sand and aggregate, such as opal, chert, and flint, or volcanics with high silica content, are reactive with the calcium, sodium, and potassium hydroxide alkalies in portland cement concrete. These reactions, though observed and studied for more than 50 years (Bureau of Reclamation, 1942), remain poorly defined and little understood. Some concrete containing alkali reactive aggregate shows immediate evidence of destructive expansion and deterioration. Other concrete might remain undisturbed for many years. Petrographic examination of reactive concrete shows that a gel is formed around the reactive aggregate. This gel undergoes extensive expansion in the presence of water or water vapor (a relative humidity of 80 to 85 percent is all the water required), creating tension cracks around the aggregate and expansion of the concrete (figure 18). If unconfined, the expansion within the concrete is first apparent by pattern cracking on the surface. Usually, some type of whitish exudation will be evident in and around the cracked concrete. In extreme instances, these cracks have opened 1.5 to 2 inches (figure 19). It is common for such expansion to cause significant offsets in the concrete and binding or seizure of control gates on dams. In large concrete structures, alkali-aggregate reaction may occur only in certain areas of the structure. Until it is recognized that multiple aggregate sources are commonly used to construct large concrete structures, this might be confusing. Only portions of the structure constructed with concrete containing alkali reactive sand and/or aggregate will exhibit expansion due to alkali-aggregate reaction. This situation presently exists at Minidoka Dam (Stark and DePuy, 1995), Stewart Mountain Dam, Coolidge Dam, Friant Dam, and Seminoe Dam.

In new construction, low alkali portland cements and fly ash pozzolan can be used to

Figure 18.—Gel resulting from alkali-aggregate reaction causes expansion and tension cracks in a concrete core.

eliminate or greatly reduce the deterioration of reactive aggregates. In existing concrete structures, deterioration due to reactive aggregate is virtually impossible to mediate. There are no proven methods of eliminating the deterioration of alkali-aggregate reaction, although the rate of expansion can sometimes be reduced by taking steps to maintain the concrete in as dry a condition as possible. It is usually futile to attempt repair of concrete actively undergoing alkali-aggregate reaction. The continuing expansion within the concrete will simply disrupt and destroy the repair material. Structures undergoing active deterioration should be monitored for rate of expansion and movement, and only the repairs necessary to maintain safe operation of the facility should be made. The binding gates of several dams have been relieved and returned to operation by using wire saws to make expansion relief cuts in the concrete on either side of the binding gates. The cuts were subsequently sealed to water leakage using polyurethane resin injection techniques (section 34). With continuing expansion of the concrete, such relief cuts may have to be repeated several times.

In many structures, the expansion and movement associated with reactive aggregate slows down and ceases when all the alkali components are consumed by the reactions. Once the expansion ceases, repairs can be performed to rehabilitate and restore the structure to full operation and serviceability. However, it should be anticipated that, ultimately, it may be necessary to replace structures undergoing alkali-aggregate deterioration. Such was the case with the 1975 replacement of Reclamation's American Falls Dam in Idaho. This dam was constructed in 1927 and replaced after extensive studies conducted by Reclamation's Denver concrete laboratories revealed that it had been severly damaged by alkali-aggregate reaction.

**16. Deterioration Caused by Cyclic Freezing and Thawing.**—Freeze-thaw deterioration is a common cause of damage to concrete constructed in the colder climates. For freeze-thaw damage to occur, the following conditions must exist:

a. The concrete must undergo cyclic freezing and thawing.

Figure 19.—Severe cracking caused by alkali-aggregate reaction.

b. The pores in the concrete, during freezing, must be nearly saturated with water (more than 90 percent of saturation).

Water experiences about 15 percent volumetric expansion during freezing. If the pores and capillaries in concrete are nearly saturated during freezing, the expansion exerts tensile forces that fracture the cement mortar matrix. This deterioration occurs from the outer surfaces inward in almost a layering manner. The rate of progression of freeze-thaw deterioration depends on the number of cycles of freezing and thawing, the degree of saturation during freezing, the porosity of the concrete, and the exposure conditions. The tops of walls exposed to snomelt or water spray, horizontal slabs exposed to water, and vertical walls at the water line are the locations most commonly damaged by freeze-thaw deterioration. If such concrete has a southern exposure, it will experience daily cycles of freezing during the night and thawing during the morning. Conversely, concrete with a northern exposure may only experience one cycle of freezing and thawing each winter, a far less damaging condition. Figures 20 and 21 show typical examples of freeze-thaw deterioration.

Another type of deterioration caused by cycles of freezing and thawing is known as D-cracking. In this instance, the expansion occurs in low quality, absorptive, coarse aggregate instead of in the cement mortar matrix. D-cracking is most commonly seen at the exposed corners of walls or slabs formed by joints. A series of roughly parallel cracks exuding calcite usually cuts across the corners of such damage (figure 22).

In 1942, Reclamation began specifying the use of air entraining admixtures (AEA) in concrete to protect concrete from freezing and thawing damage. Concrete structures built prior to that date did not contain AEA. Angostura Dam, started in 1946, was the first Reclamation Dam constructed with speci-fications requiring the use of AEA (Price, 1981). This type of admixture produces small air bubbles in the concrete matrix that provide space for water expansion during freezing. If the proper AEA, at the correct concentration, is properly mixed into high quality fresh concrete, there should be very little damage resulting from cyclic freezing and thawing except in very severe climates. Accordingly, if freezing and thawing damage is suspected in modern concrete, investigations should be performed to determine why the AEA was not effective. Except in cases of extremely cold and wet exposure, modern concrete exhibiting freeze-thaw damage has most likely suffered low durability from some other cause (see section 23).

Damage caused by cyclic freezing and thawing of concrete occurs only when the concrete is nearly saturated. Successful mitigation of freeze-thaw deterioration, therefore, involves reducing or eliminating the cycles of freezing and thawing or reducing absorption of water into the concrete. It usually is not practical to protect or insulate concrete from cycles of freezing and thawing temperatures, but concrete sealing compounds (sections 35 and 38) can be applied to exposed concrete surfaces to prevent or reduce water absorption. The sealing compounds are not effective in protecting inundated concrete, but they can provide protection to concrete exposed to rain, windblown spray, or snow melt water.

Repair of concrete damaged by freeze-thaw deterioration is most often accomplished with replacement concrete (section 29) if the damage is 6 inches or deeper, or with epoxy-bonded replacement concrete (section 31) or polymer concrete (section 32) if the damage is between 1.5 and 6 inches deep. The replacement concretes must, of course, contain AEA. Attempted repair of spalls or shallow freeze-thaw deterioration less than 1.5 inches deep is discouraged. To date, no generic or proprietary repair material tested in the Denver laboratories has been found fully suitable for such shallow repairs.

**17. Abrasion-Erosion Damage.**—Concrete structures that transport water containing silt, sand, and rock or water at high velocities are subject to abrasion damage. Dam stilling

Figure 20.—Freezing and thawing deterioration on small irrigation gate structure.

Figure 21.—Freezing and thawing deterioration on spillway concrete.

Figure 22.—D-cracking type of freezing and thawing deterioration.

basins experience abrasion damage if the flows do not sweep debris from the basins. Some stilling basins have faulty flow patterns that cause downstream sand and rock to be pulled upstream into the basins. This material is retained in the basins where it produces significant damage during periods of high flow (figure 23). Abrasion damage results from the grinding action of silt, sand, and rock. Concrete surfaces damaged in this way usually have a polished appearance (figure 24). The coarse aggregate often is exposed and somewhat polished due to the action of the silt and sand on the cement mortar matrix. Figure 25 shows an early stage of abrasion or, possibly, erosion damage to a stilling basin wall. The extent of abrasion-erosion damage is a function of so many variables—duration of exposure, shape of the concrete surfaces, flow velocity and pattern, flow direction, and aggregate loading—that it is difficult to develop general theories to predict concrete performance under these conditions. Consequently, hydraulic model studies are often required to define the flow conditions and patterns that exist in damaged basins and to evaluate required modifications. If the conditions that caused abrasion-erosion

damage are not addressed, the best repair materials will suffer damage and short service life.

It is generally understood that high quality concrete is far more resistant to abrasion damage than low quality concrete, and a number of studies (Smoak, 1991) clearly indicate that the resistance of concrete increases as the compressive strength of the concrete increases.

Abrasion damage is best repaired with silica fume concrete (section 37) or polymer concrete (section 32). These materials have shown the highest resistance to abrasion damage in laboratory and field tests. If the damage does not extend behind reinforcing steel or at least 6 inches into the concrete, the silica fume concrete should be placed over a fresh epoxy bond coat. Figure 26 shows the application of silica fume concrete to an area of abrasion, erosion, and freeze-thaw damage on the floor of the Vallecito Dam spillway.

**18. Cavitation Damage.**—Cavitation damage occurs when high velocity waterflows encounter discontinuities on the flow surface.

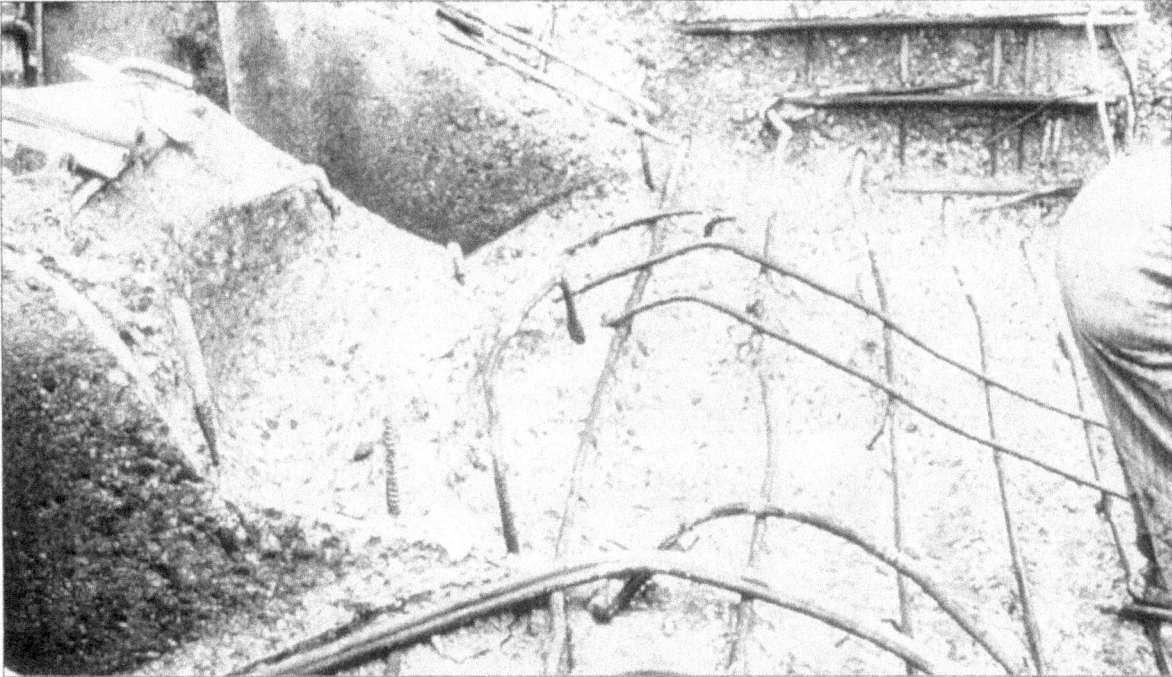

Figure 23.—Abrasion-erosion damage in a concrete stilling basin.

Figure 24.—Abrasion-erosion damage caused by sand or silt.

Figure 25.—Early stages of abrasion-erosion damage.

Discontinuities in the flow path cause the water to lift off the flow surface, creating negative pressure zones and resulting bubbles of water vapor. These bubbles travel downstream and collapse. If the bubbles collapse against a concrete surface, a zone of very high pressure impact occurs over an infinitely small area of the surface. Such high impacts can remove particles of concrete, forming another discontinuity which then can create more extensive cavitation damage. Figure 27 shows the classic "Christmas tree" pattern of cavitation damage that occurred in a large concrete-lined tunnel at Glen Canyon Dam during the flood releases of 1982. In this instance, cavitation damage extended entirely through the concrete tunnel lining and some 40 feet into foundation rock (figure 28).

Cavitation damage is common on and around water control gates and gate frames. Very high velocity flows occur when control gates are first being opened or at small gate openings. Such flows cause cavitation damage just downstream from the gates or gate frames.

The cavitation resistance of many different repair materials has been tested by the laboratories of Reclamation, the U.S. Army Corps of Engineers, and others. To date, no material, including stainless steel and cast iron, has been found capable of withstanding fully developed instances of cavitation. Successful repairs must first include mediation of the causes of cavitation.

A standard rule of thumb is that cavitation damage will not occur at flow velocities less than about 40 feet per second at ambient pressures. As flow velocities approach this threshold, it becomes necessary to ensure that there are no offsets or discontinuities on the surfaces in the flow path. Reclamation's specifications for finishing the surfaces of concrete structures that will experience high velocity flows are very strict. Repairs to newly constructed concrete that fail to meet these requirements can sometimes be accomplished by surface grinding (section 24). More likely, however, the concrete that does not meet surface specifications must be removed and

Figure 26.—Placing silica fume concrete to repair a spillway floor damaged by cyclic freezing and thawing and abrasion-erosion.

Figure 27.—Typical Christmas tree pattern of progressive cavitation damage.

Figure 28.—Extensive cavitation damage to Glen Canyon Dam.

replaced with replacement concrete (section 29) or epoxy-bonded replacement concrete (section 31).

Cavitation damage at, or adjacent to, control gates can usually be repaired with epoxy-bonded epoxy mortar (section 30), polymer concrete (section 32), or epoxy-bonded replacement concrete (section 31). Such damage is usually not very extensive in nature. That is, it is usually discovered before major repairs become necessary. After performing such repairs, it might be a good idea to apply a 100-percent solids epoxy coating to the concrete, beginning at the gate frame and extending downstream 5 to 10 feet. The glass-like surfaces of epoxy coatings may help prevent cavitation damage to the concrete. It should be understood, however, that epoxy coatings will not resist fully developed instances of cavitation damage.

Successful repair of cavitation damage to spillway, outlet works, or stilling basin concrete almost always requires making major modifications to the damaged structure to prevent recurrence of damage. Performance of hydraulic model studies should be considered to ensure correctness of the design of such repair and facility modification. One modification technique, the installation of air slots in spillways and tunnels, has been very successful in eliminating or significantly reducing cavitation damage. Replacement concrete is usually used for construction of such features and the repair of the cavitation damage.

**19. Corrosion of Reinforcing Steel.—**
Corrosion of reinforcing steel is usually a symptom of damage to concrete rather than a cause of damage. That is, some other cause weakens the concrete and allows steel corrosion to occur. However, corroded reinforcing steel is so commonly found in damaged concrete that the purposes of this guide will best be served by discussing it as if it were a cause of damage.

The alkalinity of the portland cement used in concrete normally creates a passive, basic environment (pH of about 12) around the reinforcing steel which protects it from corrosion. When that passivity is lost or destroyed, or when the concrete is cracked or delaminated sufficiently to allow free entrance of water, corrosion can occur. The iron oxides formed during steel corrosion require more space in the concrete than the original reinforcing steel. This creates tensile stresses within the concrete and results in additional cracking and/or delamination which accelerate the corrosion process.

Some of the more common causes of corrosion of reinforcing steel are cracking associated with freeze-thaw deterioration, sulfate exposure, and alkali-aggregate reaction, acid exposure, loss of alkalinity due to carbonation, lack of sufficient depth of concrete cover, and exposure to chlorides.

Exposure to chlorides greatly accelerates the rates of corrosion and can occur in several manners. The application of deicing salts (sodium chloride) to concrete to accelerate thawing of snow and ice is a common source of chlorides. Chlorides can also occur in the sand, aggregate, and mixing water used to prepare concrete mixtures. Some irrigation structures located in the Western States transport waters that have high chloride contents (figure 29). Concrete structures located in marine environments experience chloride exposure from the sea water or from windblown spray. Finally, it was once a somewhat common practice to use concrete admixtures containing chlorides to accelerate the hydration of concrete placed during winter conditions.

The occurrence of corroding reinforcing steel can usually, but not always, be detected by the presence of rust stains on the exterior surfaces and by the hollow or drummy sounds that result from tapping the affected concrete with a hammer. It can also be detected by measuring the half cell potentials of the affected concrete using special electronic devices manufactured specifically for this purpose. When the presence of corroding steel has been confirmed, it is important to define what actually caused the corrosion because the cause(s) of corrosion will usually

Figure 29.—Concrete damage caused by chloride-induced corrosion of reinforcing steel. The waters contained within this flume had high chloride content.

determine which repair procedure should be used. Further discussion of such repair procedures can be found elsewhere in this guide. Once the cause of damage has been defined and mitigated, if necessary, proper preparation of the corroded steel exposed during removal of the deteriorated concrete becomes important. Steel that has been reduced to less than half its original cross section by the corrosion process should be removed and replaced. The remaining steel must then be cleaned to remove all loose rust, scale, and corrosion byproducts that would interfere with the bond to the repair material. Corroded reinforcing steel may extend from areas of obviously deteriorated concrete well into areas of apparently sound concrete. Care must be taken to remove sufficient concrete to include all the corroded steel.

**20. Acid Exposure.**—The more common sources of acidic exposure involving concrete structures occur in the vicinity of underground mines. Drainage waters exiting from such mines can contain acids of sometimes unexpectedly low pH value. A pH value of 7 is defined as neutral. Values higher than 7 are

defined as basic, while pH values lower than 7 are acidic. A 15- to 20-percent solution of sulfuric acid will have a pH value of about 1. Such a solution will damage concrete very rapidly. Acidic waters having pH values of 5 to 6 will also damage concrete, but only after long exposure.

Concrete damaged by acids is very easy to detect. The acid reacts with the portland cement mortar matrix of concrete and converts the cement into calcium salts that slough off or are washed away by flowing waters. The coarse aggregate is usually undamaged but left exposed. The appearance of acid-damaged concrete is somewhat like that of abrasion damage, but the exposure of the coarse aggregate is more pronounced and does not appear polished. Figures 30 and 31 show the typical appearance of concrete that has been damaged by acid exposure. Acid damage begins, and is most pronounced, on the exposed surface of concrete but always extends, to a diminishing extent, into the core of the structure. The acid is most concentrated at the surface. As it penetrates into the concrete, it is neutralized by reaction with the

Figure 30.—Concrete deterioration caused by acidic water.

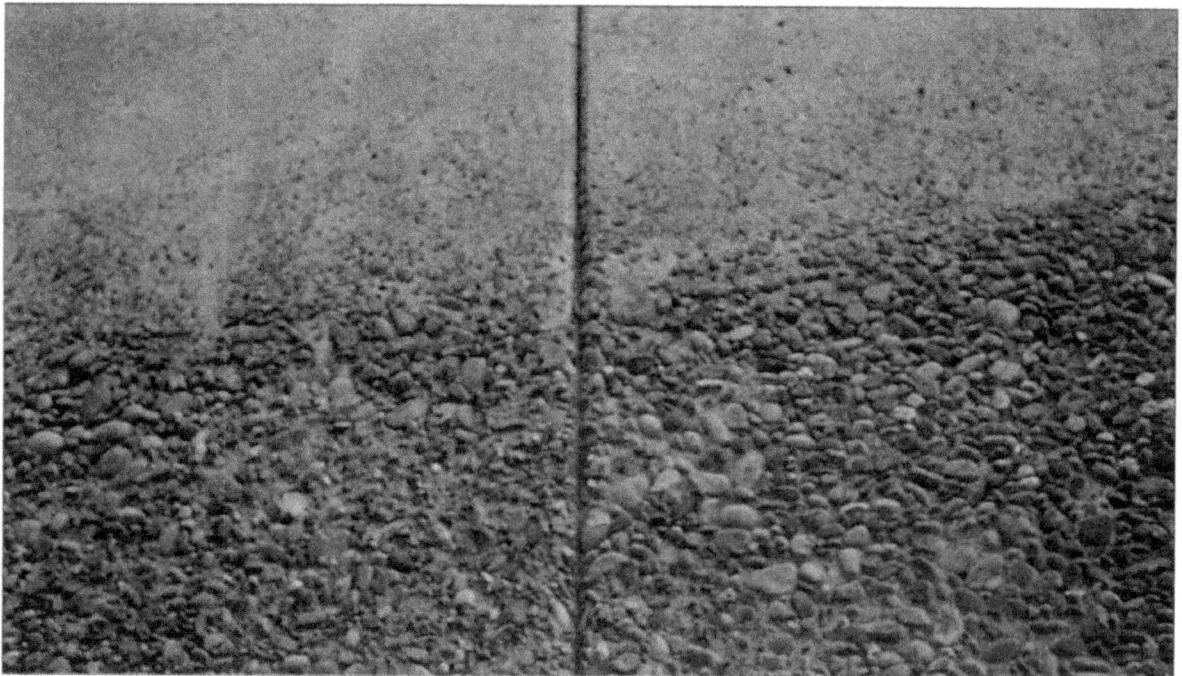

Figure 31.—The depth of acidic water on this concrete wall is very apparent.

portland cement. The cement at depth inside the structure, however, is weakened by the reaction. Preparation of acid-damaged concrete, therefore, always involves removal of more concrete than would otherwise be expected. Failure to remove all the concrete affected and weakened by the acid will result in bond failure of the repair material. Acid washes were once permitted as a method of cleaning concrete surfaces in preparation for repairs. It has been learned, however, that bond failures would occur unless extensive efforts were expended to remove all traces of acid from the concrete. Reclamation specifications no longer permit the use of acid washes to prepare concrete for repair or to clean cracks subject to resin injection repairs.

As with all causes of damage to concrete, it is generally necessary to remove the source of damage prior to repair. The most common technique used with acid damage is to dilute the acid with water. Low pH acid solutions can be converted to higher pH solutions having far less potential for damage in this manner. Alternately, if the pH of the acid solution is relatively high, coatings such as the thin polymer concrete coating system, section 33, can be applied over repair materials to prevent the acid from redamaging the surfaces. Laboratory tests have revealed very few economical coatings capable of protecting repair materials from low pH solutions.

Repairs to acid-damaged concrete can be made using epoxy-bonded replacement concrete (section 31), replacement concrete (section 29), polymer concrete (section 32), and, in some instances, epoxy-bonded epoxy mortar (section 30). Polymer concrete and epoxy mortar, which do not contain portland cement, offer the most resistance to acid exposure conditions.

**21. Cracking.**—Cracking, like corrosion of reinforcing steel, is not commonly a cause of damage to concrete. Instead, cracking is a symptom of damage created by some other cause.

All portland cement concrete undergoes some degree of shrinkage during hydration. This shrinkage produces multidirectional drying shrinkage and curing shrinkage cracking having a somewhat circular pattern (figure 32). Such cracks seldom extend very deeply into the concrete and can generally be ignored.

Plastic shrinkage cracking occurs when the fresh concrete is exposed to high rates of evaporative water loss which causes shrinkage while the concrete is still plastic (figure 33). Plastic shrinkage cracks are usually somewhat deeper than drying or curing shrinkage cracks and may exhibit a parallel orientation that is visually unattractive.

Thermal cracking is caused by the normal expansion and contraction of concrete during changes of ambient temperature. Concrete has a linear coefficient of thermal expansion of about 5.5 millionths inch per inch per degree Fahrenheit (°F). This can cause concrete to undergo length changes of about 0.5 inch per 100 linear feet for an 80 °F temperature change. If sufficient joints are not provided by the design of the structure to accommodate this length change, the concrete will simply crack and provide the joints where needed. This type of cracking will normally extend entirely through the member and create a source of leakage in water retaining structures. Thermal cracking can also be caused by using portland cements developing high heats of hydration during curing. Such concrete develops exothermic heat and hardens while at elevated temperatures. Subsequent contraction upon cooling develops internal tensile stresses and resulting cracks at or across points of restraint.

Inadequate foundation support is another common cause of cracking in concrete structures. The tensile strength of concrete is usually only about 200 to 300 psi. Foundation settlement can easily create displacement conditions where the tensile strength of concrete is exceeded with resulting cracking.

Cracking is also caused by alkali-aggregate reaction, sulfate exposure, and exposure to cyclic freeze-thaw conditions, as has been

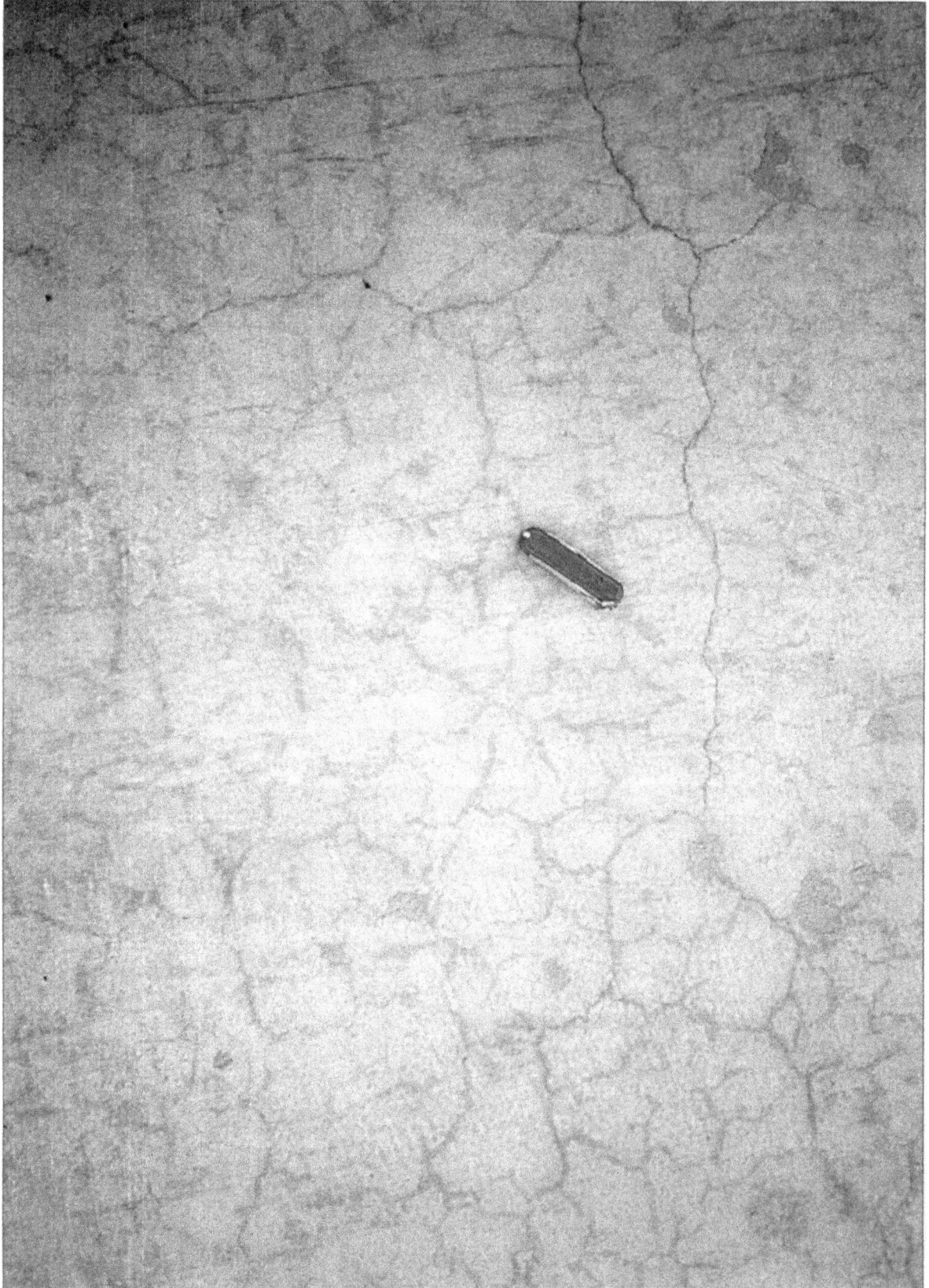

Figure 32.—Typical appearance of drying shrinkage cracking.

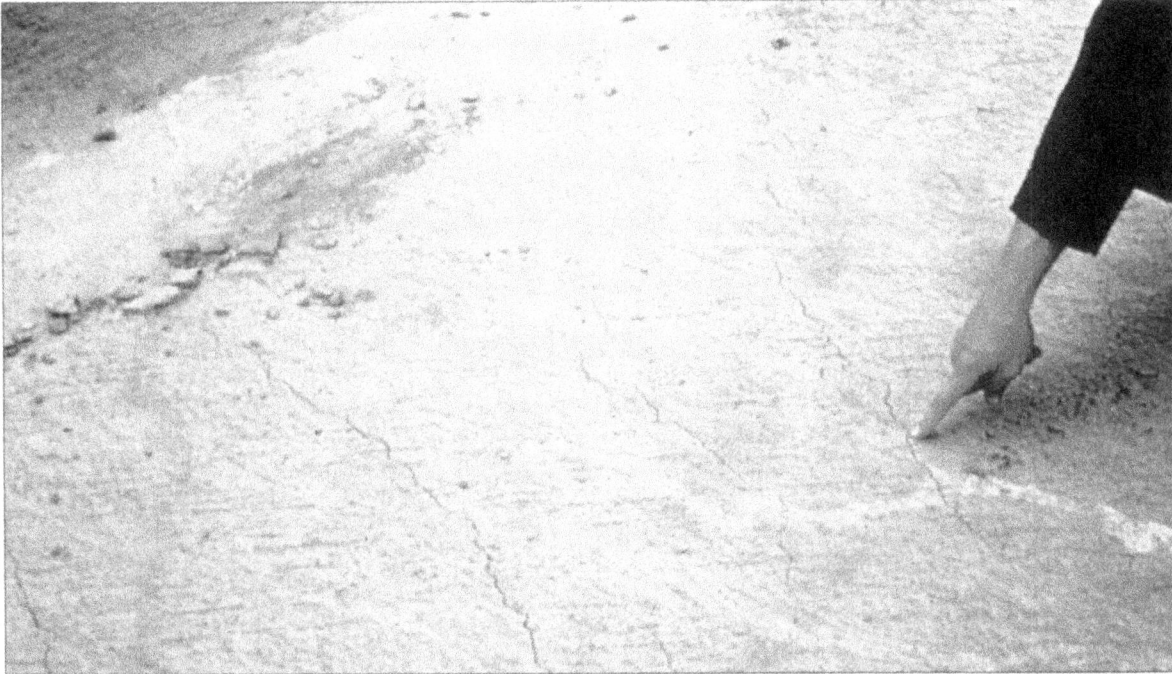

Figure 33.—Plastic shrinkage cracking caused by high evaporative water loss while the concrete was still in a plastic state.

discussed in previous sections, and by structural overloads as discussed in the following section.

Successful repair of cracking is often very difficult to attain. It is better to leave most types of concrete cracking unrepaired than to attempt inadequate or improper repairs (figure 34 and 35). The selection of methods for repairing cracked concrete depends on the cause of the cracking. First, it is necessary to determine if the cracks are "live" or "dead." If the cracks are cyclicly opening and closing, or progressively widening, structural repair becomes very complicated and is often futile. Such cracking will simply reestablish in the repair material or adjacent concrete. For this reason, it is normal procedure to install crack gages across the cracks to monitor their movement prior to attempting repair (figure 36). The gages should be monitored for extended periods to determine if the cracks are simply opening and closing as a result of daily or seasonal temperature changes or if there is a continued or progressive widening of the cracks resulting from foundation or load conditions. Repairs should be attempted only after the cause and behavior of the cracking is understood.

If it is determined that the cracks are "dead" or static, epoxy resin injection (section 34) can be used to structurally rebond the concrete. If the objective of the repair is to seal water leakage rather than to accomplish structural rebonding, the cracks should be injected with polyurethane resin. Epoxy resin injection can sometimes be used to seal low volume water leakage and structurally rebond cracked concrete members. Epoxy resins cure to form hard, brittle materials that will not withstand movement of the injected cracks. Polyurethane resins cure to a flexible, low tensile strength, closed cell foam that is effective in sealing water leakage but cannot normally be used for structural rebonding. (Some two component polyurethane resin systems cure to form flexible solids that may be useful for structural rebonding.) These flexible foams can experience 300- to 400-percent elongation due to crack movement.

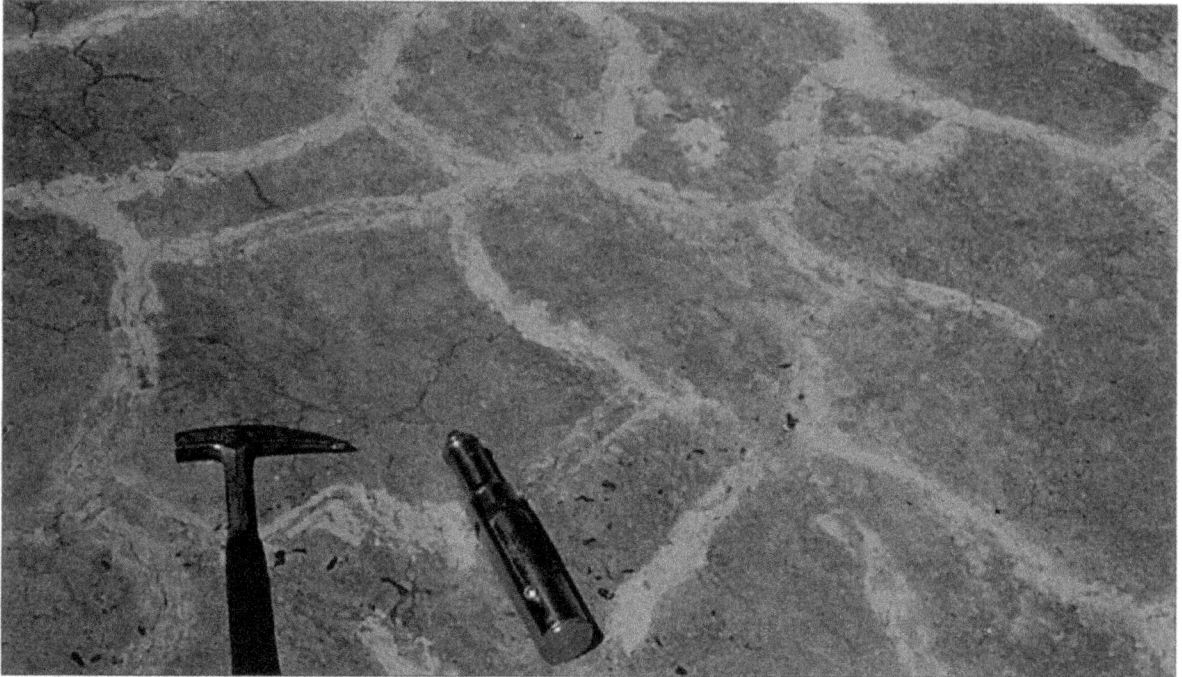

Figure 34.—Inadequate crack repair techniques often result in poor appearance upon completion.

Figure 35.—Improper crack repair techniques often result in short service life.

Figure 36.—Crack gage installed across a crack will allow determination of progressive widening or movement of the crack. It may be necessary to monitor such gages for periods up to a year to predict future crack behavior.

It is not uncommon to find that damaged concrete contains cracking not related to the cause of the primary damage (see section 23). If the depth of removal of the damaged or deteriorated concrete does not extend below the depth and extent of the existing cracks, it should be expected that the cracking will ultimately reflect through the new repair materials. Such reflective cracking is common in bonded overlay repairs to bridge decks, spillways, and canal linings (figure 37). If reflective cracking is intolerable, the repairs must be designed as separate structural members not bonded to the old existing concrete.

**22. Structural Overloads.**—Concrete damage caused by structural overloading is usually very obvious and easy to detect. Frequently, the event causing overloading has been noted and is a matter of record. The stresses created by overloads result in distinctive patterns of cracking that indicate the source and cause of excessive loading and the point(s) of load application. Normally, structural overloads

are one time events and, once defined, the resulting damage can be repaired with the expectation that the cause of the damage will not reoccur to create damage in the repaired concrete.

It should be expected that the assistance of a knowledgeable structural engineer will be required to perform the structural analysis needed to fully define and evaluate the cause and resulting damage of most structural overloads and to assist in determining the extent of repair required. This analysis should include determination of the loads the structure was designed to carry and the extent the overload exceeded design capacity. A thorough inspection of the damaged concrete must be performed to determine the entire effect of the overload on the structure. Displacements must be discovered and the secondary damage, if any, located. Care should be taken to ensure that some other cause of damage did not first weaken the concrete and make it incapable of carrying the design loads.

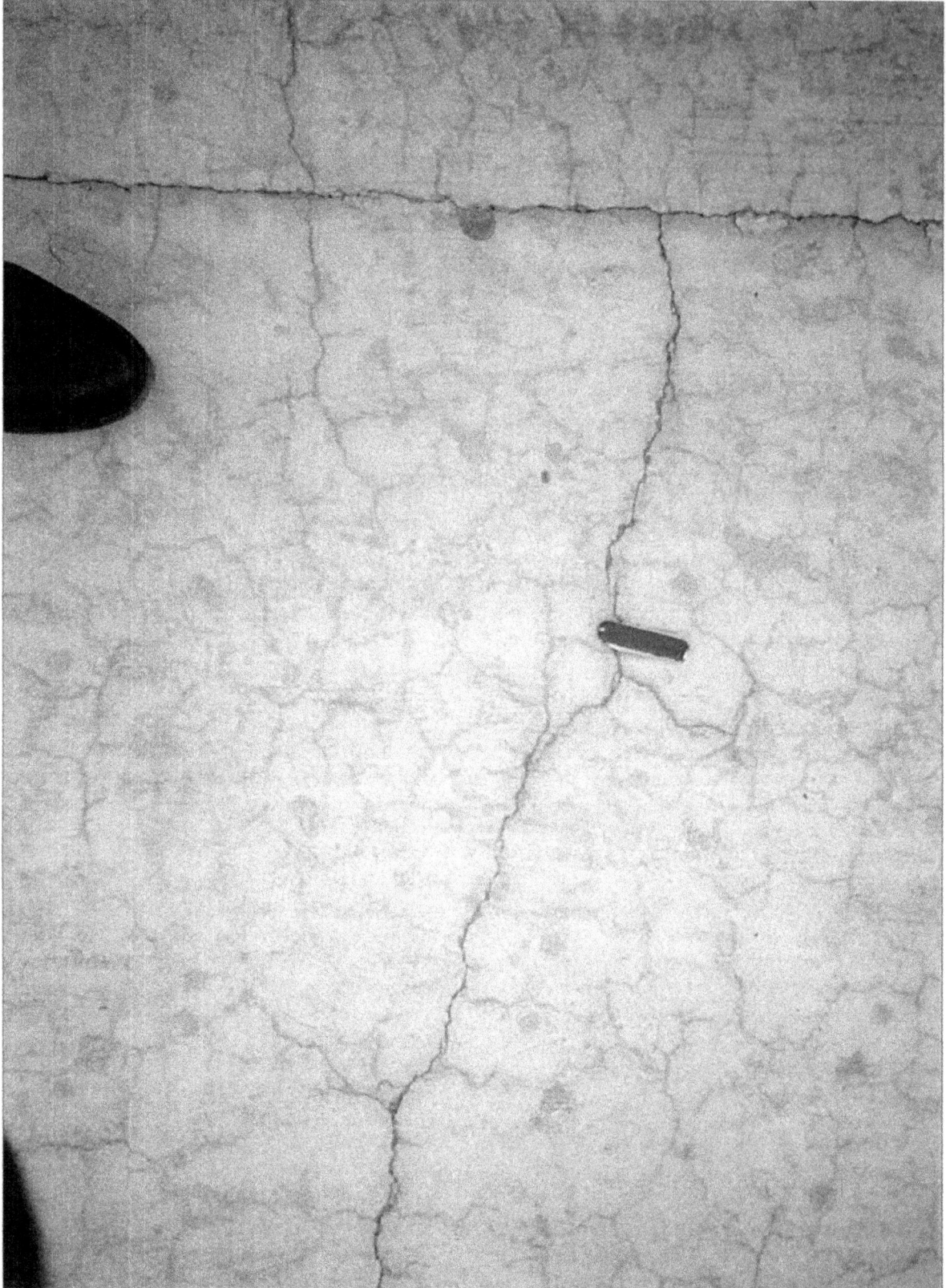

Figure 37.—A large reflective crack has formed in a concrete overlay which also exhibits circular drying shrinkage cracking.

The repair of damage caused by overloading can, most likely, best be performed with conventional replacement concrete (section 29). The need for repair and/or replacement of damaged reinforcing steel should be anticipated and included in the repair procedures.

**23. Multiple Causes of Damage.**—Multiple causes of the damage should be suspected whenever damage or deterioration is discovered in modern concrete. Modern concrete (concrete constructed since about 1950) has the benefit of the use of various admixtures and advanced concrete materials technology. Such concrete should not be damaged by many of the causes listed in this chapter. If deterioration or damage has occurred, it is likely that a combination of causes are in effect. Failure to recognize and mitigate all the causes of damage will most likely result in poor repair serviceability. Figure 38 shows the results of multiple cause damage. This concrete is suffering from alkali-aggregate reaction cracking that has also accelerated freeze-thaw deterioration of the surface. It is also being damaged by faulty design or construction techniques that located the electrical conduits too close to the exterior surface of the concrete.

The proper use of air entraining admixtures in modern concrete has greatly increased the resistance of the concrete to freeze-thaw deterioration. Except in unusually severe exposures, freeze-thaw deterioration should not occur. This notwithstanding, freeze-thaw deterioration is often still blamed as the cause of damage to modern concrete. Before blaming freezing and thawing conditions, it is better to first determine why the air entraining admixture did not provide effective protection. Mix records and aggregate quality test results may indicate that the concrete was poorly proportioned or that the available aggregate was of low quality. Construction inspection records may indicate that placing and finishing techniques were inadequate. Petrographic examination of the affected concrete may reveal that alkali-aggregate reaction, sulfate exposure, or induced chlorides have weakened the concrete and allowed freeze-thaw damage to occur. All such findings might indicate that the problem is far more extensive than at first thought and requires more extensive preventative or corrective action than the simple replacement of the presently deteriorated concrete. The use of excessive mix water, the improper type of portland cement, poor construction practices, improper mixture proportioning, dirty or low quality aggregate, and inadequate curing all contribute to low durability in concrete. Such concrete may have low resistance to normal weathering or to other hazards.

Selection of the proper methods and materials for repair of concrete damaged by multiple causes depends on determining which is the weakening cause and which is the accelerated cause. Once the weakening cause is fully understood, it is commonly necessary to take preventative measures to protect the remaining original concrete from additional damage. The application of concrete sealing compounds (sections 35 and 38) or the thin polymer concrete overlay (section 32) may prove useful in this respect. If no such preventative measures are judged useable, repair of the damaged concrete can be made as discussed in previous sections, but short repair service life and the occurrence of future damage should be anticipated.

Figure 38.—Multiple causes of damage are apparent in this photograph. Poor design or construction practices placed the electrical conduit too near the surface. A combination of freezing and thawing deterioration and alkali-aggregate reaction is responsible for the cracking and surface spalling on the parapet wall.

# Chapter IV

# Standard Methods of Concrete Repair

Proven methods of repairing concrete are described in this chapter. Sections 24 through 38 contain detailed discussions of each of the proven repair methods. Construction specifications for these methods/materials of repair are contained in the latest revision of Reclamation's *Standard Specifications for the Repair of Concrete, M-47*, Appendix A. It is essential that the provisions of these specifications be closely followed during repair of Reclamation concrete. It should be recognized, however, that these "standard" methods and specifications cannot apply to unusual or nonstandard concrete repair situations. Assistance with unusual or special repair problems can readily be obtained by contacting personnel of the Materials Engineering and Research Laboratories, Code D-8180.

**24. Surface Grinding.**—Surface grinding can be used to repair some bulges, offsets, and other irregularities that exceed the desired surface tolerances. Excessive surface grinding, however, may result in weakening of the concrete surface, exposure of easily removed aggregate particles, or unsightly appearance. For these reasons, surface grinding should be performed subject to the following limitations:

   a. Grinding of surfaces subject to cavitation erosion (hydraulic surfaces subject to flow velocities exceeding 40 feet per second) should be limited in depth so that no aggregate particles more than 1/16 inch in cross section are exposed at the finished surface.

   b. Grinding of surfaces exposed to public view should be limited in depth so that no aggregate particles more than 1/4 inch in cross section are exposed at the finished surface.

   c. In no event should surface grinding result in exposure of aggregate of more than one-half the diameter of the maximum size aggregate.

Where surface grinding has caused or will cause exposure of aggregate particles greater than the limits of subparagraph 24.a. or b., the concrete must then be repaired by excavating and replacing the concrete in accordance with sections 29, 30, 31, or 32.

**25. Portland Cement Mortar.**—Portland cement mortar may be used for repairing defects on surfaces not prominently exposed, where the defects are too wide for dry pack filling or where the defects are too shallow for concrete filling and no deeper than the far side of the reinforcement that is nearest the surface. Repairs may be made either by use of shotcrete or by hand application methods. Replacement mortar can be used to make shallow, small size repairs to new or green concrete, provided that *the repairs are performed within 24 hours of removing the concrete forms.*

The use of replacement mortar to repair old or deteriorated concrete has been permitted on Reclamation projects in the past. It is now recognized, however, that accomplishing successful mortar repairs to old concrete without the use of a bonding resin is unlikely or extremely difficult. Evaporative loss of water from the surface of the repair mortar, combined with capillary water loss to the old concrete, results in unhydrated or poorly hydrated cement in the mortar. Additionally, repair mortar bond strength development proceeds at a slower rate than compressive strength development. This causes workers to mistakenly abandon curing procedures prematurely, when the mortar "seems strong."

Once the mortar dries, bond strength development stops, and bond failure of the mortar patch results.

For these reasons, using cement mortar without a resin bond coat to repair old concrete is discouraged and is not allowed under Reclamation's *Standard Specifications for the Repair of Concrete, M-47*.

A portland cement mortar patch is usually darker than the surrounding concrete unless precautions are taken to match colors (figure 39). A leaner mix will usually produce a lighter color patch. Also, white cement can be used to produce a patch that will blend with the surrounding concrete. The quantity of white cement to use must be determined by trial.

*(a) Preparation.*—Concrete to be repaired with replacement mortar should be prepared in accordance with the provisions of section 8. After preparation, the areas should be cleaned, roughened, if necessary (preferably by wet sandblasting), and surface dried to a saturated surface condition. The mortar should be applied immediately thereafter.

*(b) Materials.*—Replacement mortar contains water, portland cement, and sand. The cement should be same type as used in the concrete being repaired. The water and sand should be suitable for use in concrete, and the sand should pass a No. 16 sieve. The cement to sand ratio should be between 1:2 to 1:4, depending on application technique. Only enough water should be added to the cement-sand mixture to permit placing.

*(c) Application.*—Best results with replacement mortar are obtained when the material is pneumatically applied using a small gun. Equipment commonly used for shotcreting is too large to be satisfactory for the ordinarily small size mortar repairs of new concrete. With shotcreting equipment, neat work is difficult in the usual small areas, and cleanup costs are high because cleanup is seldom done promptly. However, small size equipment such as shown in figure 40 has been

satisfactory for small scale repair work when the mortar was premixed, including water, to a consistency of dry-pack material. No initial application of cement, cement grout, or wet mortar should be made. If repairs are more than 1 inch deep, the mortar should be applied in layers not more than 3/4 of an inch thick to avoid sagging and loss of bond. After completion of each layer, there should be a lapse of 30 minutes or more before the next layer is placed. Scratching or otherwise preparing the surface of a layer prior to addition of the next layer is unnecessary, but the mortar must not be allowed to dry.

In completing the repair, the hole should be filled slightly more than level full. After the material has partially hardened but can still be trimmed off with the edge of a steel trowel, excess material should be shaved off, working from the center toward the edges. Extreme care must be used to avoid impairment of bond. Neither the trowel nor water should be used in finishing. A satisfactory finish may be obtained by lightly rubbing the surface with a soft rag.

For minor restorations, satisfactory mortar replacement may be performed by hand. The success of this method depends on complete removal of all defective and affected concrete, good bonding of the mortar to the concrete, elimination of shrinkage of the patch after placement, and thorough curing.

Replacement mortar repairs can be made using an epoxy bonding agent as described in section 26. This technique, while not required by Reclamation's M-47 specifications, is highly recommended.

*(d) Curing.*—Failure to cure properly is the most common cause of failure of replacement mortar. It is essential that mortar repairs receive a thorough water cure starting immediately after initial set and continuing for 14 days. In no event should the mortar be allowed to become dry during the 14-day period following placement. Following the 14-day water cure and while the mortar is still saturated, the surface of the mortar should

Figure 39.—A portland cement mortar patch seldom matches the color of the original concrete unless special efforts are taken to blend white cement with normal portland cement.

Figure 40.—A small size pneumatic gun can be used to apply portland cement mortar. Regular shotcreting equipment would be too large for this application.

be coated with two coats of a wax-base or water-emulsified resin base curing compound meeting Reclamation specifications. If this curing procedure cannot be followed or if conditions at the job are such that this curing procedure will not be followed, money would be saved by using another repair material.

## 26. Dry Pack and Epoxy-Bonded Dry Pack.—Dry pack is a combination of portland cement and sand passing a No. 16 sieve mixed with just enough water to hydrate the cement. Dry pack should be used for filling holes having a depth equal to, or greater than, the least surface dimension of the repair area; for cone bolt, she bolt, core holes, and grout-insert holes; for holes left by the removal of form ties; and for narrow slots cut for repair of cracks. Dry pack should not be used for relatively shallow depressions where lateral restraint cannot be obtained, for filling behind reinforcement, or for filling holes that extend completely through a concrete section.

For the dry pack method of concrete repair, holes should be sharp and square at the surface edges, but corners within the holes should be rounded, especially when water tightness is required. The interior surfaces of holes left by cone bolts and she bolts should be roughened to develop an effective bond; this can be done with a rough stub of 7/8-inch steel-wire rope, a notched tapered reamer, or a star drill. Other holes should be undercut slightly in several places around the perimeter, as shown in figure 41. Holes for dry pack should have a minimum depth of 1 inch.

*(a) Preparation.*—Application of dry pack mortar should be preceded by a careful inspection to see that the hole is thoroughly cleaned and free from mechanically held loose pieces of aggregate. One of the three following methods should be used to ensure good bond of the dry pack repair.

The first method is the application of a stiff mortar or grout bond coat immediately before applying the dry pack mortar. The mix for the bonding grout is 1:1 cement and fine sand mixed with water to a fluid paste consistency.

All surfaces of the hole are thoroughly brushed with the grout, and dry packing is done quickly before the bonding grout can dry. Under no circumstances should the bonding coat be so wet or applied so heavily that the dry pack material becomes more than slightly rubbery. When a grout bond coat is used, the hole to be repaired can be dry. Presoaking the hole overnight with wet rags or burlap prior to dry packing may sometimes give better results by reducing the loss of hydration water, but there must be no free surface water in the hole when the bonding grout is applied.

The second method of ensuring good bond starts with presoaking the hole overnight with wet rags or burlap. The hole is left slightly wet with a small amount of free water on the inside surfaces. The surfaces are then dusted lightly and slowly with cement using a small dry brush until all surfaces have been covered and the free water absorbed. Any dry cement in the hole should be removed using a jet of air before packing begins. The hole should not be painted with neat cement grout because it could make the dry pack material too wet and because high shrinkage would prevent development of the bond that is essential to a good repair.

A third method of ensuring good bond is the use of an epoxy bonding resin. The epoxy bonding resin should meet the requirements of ASTM C-881 for a type II, grade 2, class B or C resin, depending on the job site ambient temperatures. Epoxies bond best to dry concrete. It may be necessary to dry the hole immediately prior to dry packing using hot air, a propane torch, or other appropriate method. The concrete temperature, however, should not be high enough to cause instant setting of the epoxy or to burn the epoxy when it is applied. After being mixed, the epoxy is thoroughly brushed to cover all surfaces, but any excess epoxy is removed. Dry pack mortar is then applied immediately, before the epoxy starts to harden. The epoxy must be either fluid or tacky when dry packing takes place. If it appears that the epoxy may become hard before dry packing is

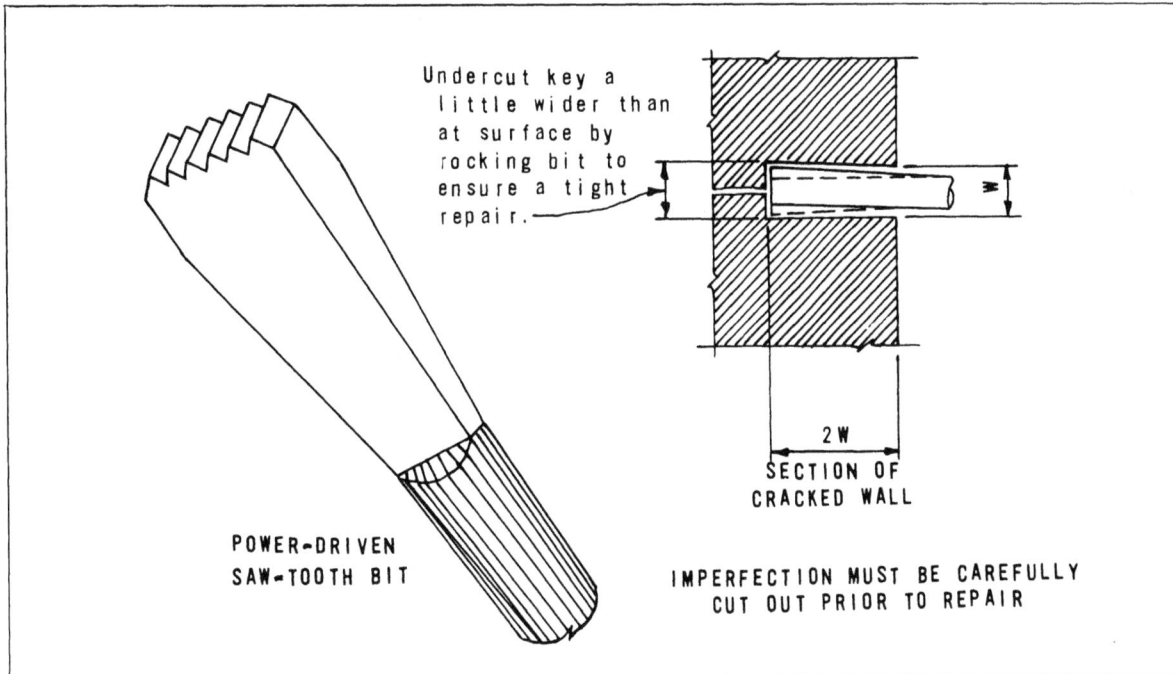

Figure 41.—Saw-tooth bit used to cut slot for dry packing.

complete, fresh fluid epoxy can be brushed over epoxy that has become tacky. If the epoxy becomes hard, it must be removed before a new coat is applied. The epoxy ensures a good bond between the dry pack repair and the old concrete. It also reduces the loss of hydration water from the repair to the surrounding concrete, thus assisting in good curing; however, the epoxy-bonded dry pack still requires curing as discussed below. Where appearance is not important, epoxy has sometimes been used on the surface in place of a curing compound. This procedure is not recommended and is not allowed on Reclamation jobs.

*(b) Materials.*—Dry pack mortar is usually a mixture (by dry volume or weight) of 1 part cement to 2-1/2 parts sand that will pass a No. 16 screen. While the mixture is rich in cement, the low water content prevents excessive shrinkage and gives high strengths. A dry pack repair is usually darker than the surrounding concrete unless special precautions are taken to match the colors. Where uniform color is important, white cement may be used in sufficient amount (as determined by trial) to produce uniform appearance. For packing cone bolt holes, a leaner mix of 1:3 or 1:3-1/2 will be sufficiently strong and will blend better with the color of the wall. Sufficient water should be used to produce a mortar that will just stick together while being molded into a ball with the hands and will not exude water but will leave the hands damp. The proper amount of water will produce a mix at the point of becoming rubbery when solidly packed. Any less water will not make a sound, solid pack; any more will result in excessive shrinkage and a loose repair.

*(c) Application.*—Dry pack mortar should be placed and packed in layers having a compacted thickness of about three-eighths of an inch. Thicker layers will not be well compacted at the bottom. The surface of each layer should be scratched to facilitate bonding with the next layer. One layer may be placed immediately after another unless an appreciable rubbery quality develops; if this occurs, work on the repair should be delayed 30 to 40 minutes. Under no circumstances should alternate layers of wet and dry materials be used.

Each layer should be solidly compacted over the entire surface by striking a hardwood dowel or stick with a hammer. These sticks are usually 8 to 12 inches long and not over 1 inch in diameter and are used on fresh mortar like a caulking tool. Hardwood sticks are used in preference to metal bars because the latter tend to polish the surface of each layer and, thus, make bonding less certain and filling less uniform. Much of the tamping should be directed at a slight angle and toward the sides of the hole to ensure maximum compaction in these areas. The holes should not be overfilled; finishing may usually be completed at once by laying the flat side of a hardwood piece against the fill and striking it several good blows with a hammer. If necessary later, a few light strokes with a rag may improve appearance. Steel finishing tools should not be used, and water must not be used to facilitate finishing.

*(d) Curing and Protection.*—Procedures for curing and protection of dry pack are essentially the same as those for concrete and are described in section 25. Additionally, the dry pack repair area should be protected and not exposed to freezing temperatures for at least 3 days after application of the curing compound.

## 27. Preplaced Aggregate Concrete.—
Preplaced aggregate concrete is an excellent repair material that has not been used much in recent years. Preplaced aggregate concrete is made by injecting portland cement grout, with or without sand, into the voids of a formed, compacted mass of clean, graded, coarse aggregate. The preplaced aggregated is washed and screened to remove fines before placing into the forms. As the grout is injected or pumped into the forms, it displaces any included air or water and fills the voids around the aggregate, thus creating a dense concrete having a high aggregate content.

Because the coarse aggregate has point contact prior to grout injection, preplaced aggregate concrete undergoes very little settlement, curing, or drying shrinkage during hydration. Drying shrinkage of preplaced aggregate concrete containing 1-1/2 inch maximum size aggregate is about 200 to 400 millionths, while conventional concrete drying shrinkage containing the same size maximum aggregate is about 400 to 600 millionths.

Another advantage of preplaced aggregate concrete is the ease with which it can be placed in certain situations where placement of conventional concrete would be extremely difficult or impossible. Preplaced aggregate concrete is especially useful in underwater repair construction. It has been used in a variety of large concrete and masonry repairs, including bridge piers and the resurfacing of dams. It has been used to construct atomic reactor shielding and plugs for outlet works and tunnels in mine workings, and it has been used to embed penstocks and turbine scrollcases (American Concrete Institute, 1992). Figure 42 shows the upstream face of Barker Dam near Boulder, Colorado, which was resurfaced with prepacked aggregate concrete.

Although preplaced aggregated is adaptable to many special repair applications, it is essential that the work be undertaken by well qualified personnel who are willing to follow exactly the construction procedures required for this repair material. Form work for preplaced aggregate concrete requires special attention to prevent grout loss. The construction of forms should be with workmanship better than that normally encountered with conventional concrete. Leaking forms can cause significant problems and should, by careful construction, be avoided whenever possible. The injected grout is more flowable than plastic concrete and takes slightly longer to set. Forms, therefore, must be constructed to take more lateral pressure than would be necessary with conventional concrete. Form bolts should fit tightly through the sheathing, and all possible points of grout leakage should be caulked.

*(a) Preparation.*—The preparation of concrete to be repaired by preplaced aggregate concrete is identical to the preparation required for replacement concrete (section 29) if the development of bond is required.

Figure 42.—The downstream face of Barker Dam, near Boulder, Colorado, was resurfaced with prepacked aggregate.

*(b) Materials.*—Grout for preplaced aggregate concrete may be mixed with sand either of the gradation specified for conventional concrete or with fine sand, pozzolanic or fly ash fillers, water reducing admixtures, and pumping admixtures as dictated by the minimum size of the coarse aggregate. With 1-1/2-inch minimum size coarse aggregate, the sand gradation is that specified for conventional concrete. The portland cement, water, and sand are mixed using high speed centrifugal grout mixers that produce well mixed grouts of a creamy consistency. For use with 1/2-inch minimum size coarse aggregate, a grout mixture is prepared containing fine sand passing a No. 8 screen and with at least 95 percent passing a No. 16 screen. Best pumping characteristics will be obtained with fineness modulus between 1.2 and 2 and with the rounded shape of natural sands as opposed to crushed sands.

Addition of fly ash and water reducing admixture improves the flowability of the grout and the ultimate strength. Proprietary pumping admixtures are commonly used to increase the penetration and pumpability of the final grout. The consistency of grout for preplaced aggregated should be uniform from batch to batch and should be such that it can be readily pumped into the voids at relatively low pressure. Consistency is affected by water content, sand grading, filler type and content, cement type, and admixture type. For each mix, there are optimum proportions that produce best grout pumpability or consistency, and tests are necessary for each job to determine these optimum proportions.

The maximum size coarse aggregate used with both types of grout is the largest available, provided that the aggregate can be easily handled and placed. Coarse aggregate should meet all the requirements of coarse aggregate for conventional concrete. It is essential that the coarse aggregated be clean. The aggregate should be well graded from minimum size (1/2-inch minimum or 1-1/2-inch minimum) up to the maximum size, and when compacted into the forms, should have a void content of 35 to 40 percent. If grout containing sand of concrete grading is used, the minimum coarse aggregate size should be 1-1/2 inches.

*(c) Application.*—The grout piping system used with preplaced aggregate concrete must be designed to serve at least 3 purposes—to deliver and inject grout, to provide means for determining grout level in the forms, and to serve as vents in enclosed forms for escape or air and water. Proper design and location of the grout piping system is essential for successful placement.

The grout delivery pipeline should be a recirculating system. That is, the grout delivery pipeline should extend from the grout agitator or holding tank to the grout pump, then to the injection manifold, and return to the grout agitator tank. With this type of pipeline, the grout can be kept moving and circulating in the delivery pipeline even when no grout is being injected into the aggregate. Such a system prevents stoppages and clogging of the delivery line. Noncirculating or deadheaded grout delivery lines are not allowed on Reclamation projects. The delivery line should be kept as short as practicable, and the pipe size should be such that normal grout flow velocities range between 2 and 4 feet per second. For most applications, a 1-inch ID grout line will suffice. All valves used in the grout piping system should be quick opening ball valves which can be readily cleaned.

The simplest piping system is a single recirculating delivery line attached via a manifold and valves to a single injection line. The injection line should extend to the lowest point in the form. Multiple injection lines are used for larger projects. Spacing of the injection lines is variable, depending on the form configuration, aggregate gradation, and other factors, but spacings of 4 to 6 feet are common. In preparing the layout of the grout delivery system, it is normally assumed that the slope of the grout face will be 4:1 for work in the dry and 6:1 for underwater work. Much flatter slopes are common with actual grout surfaces.

Sounding wells constructed from 2-inch-diameter slotted pipe are installed to allow determination of the level of grout during injection. Similarly, clear plastic windows can be installed in the forms to allow visual determination of grout levels. The number and location of sounding wells are determined by the size and configuration of the aggregate mass. The ratio of sounding wells to injection pipes should be from 1:4 to about 1:8.

Grout injection should begin at the lowest point of the form and continue uniformly until the entire form is filled. After sufficient grout has been pumped to raise the level of grout in the form about 18 inches above the bottom outlet of the injection line, the injection line can be progressively raised, maintaining about 12 inches of embedment below the level of the grout at all times. A great deal of thought and planning is required if multiple injection lines are used. The objective is to entirely fill the form without trapping air or water. Vents must be located where needed and the injection sequence designed to promote complete filling. It is not possible to use internal vibrators to consolidate preplaced aggregate concrete. External vibrators, however, can be attached to the forms and used advantageously. External vibration will eliminate the splotchy appearance that can occur where coarse aggregate particles contact the forms. Underwater applications of preplaced aggregate concrete require additional considerations. During injection, grout pumping must continue until an undiluted flow of grout emerges from the top of the form. Formwork is usually closed at the top to prevent washout or dilution of the grout after placement if flowing water is encountered. Anti-washout admixtures might prove useful for underwater applications of preplaced aggregate concrete. Care must be taken, however, when using several different types of admixtures (e.g., anti-washout, pumping aids, or high range water reducers) that undesirable combinations are avoided. It is known for example, that some anti-washout admixtures can significantly reduce the pumpability benefits of some high range water reducers. Such problems should be detected during the mixture proportioning tests previously recommended in paragraph (b).

The minimum volume of the grout mixer tank and the grout agitator tank should be 17 cubic feet. The grout should be mixed using a high speed centrifugal mixer operating at a minimum of 1,500 rotations per minute. The grout pump should be of the helical screw, rotor-type (commonly known as a "Moyno" grout pump), capable of pumping at least 20 gallons of grout per minute at the specified injection pressure.

Quality control of preplaced aggregate concrete lies with proper compaction of the aggregate into the forms and maintenance of proper grout consistency throughout the job. Compaction requirements must be satisfied by visual inspection during placement and before grout is introduced into the forms. Grout consistency can be determined by using a Baroid Model 140 Mud Balance to measure grout density. Some practitioners promote using a flow cone to time the rate of flow of a known volume of grout through the cone as a measure of consistency. Recent laboratory tests (Smoak, 1993), however, have proven that the flow cone is useless for measuring the consistency of grout containing high range water reducing admixture.

*(d) Curing.*—The curing requirements for preplaced aggregate concrete are the same as for replacement concrete (section 29). Preplaced aggregate concrete placed during underwater applications will normally receive excellent curing without further effort.

**28. Shotcrete.**—Shotcrete is defined as "mortar or concrete pneumatically projected at high speed onto a surface" (American Concrete Institute, 1990). There are two basic types of shotcrete—dry mix and wet mix. In dry mix shotcrete, the dry cement, sand, and coarse aggregate, if used, are premixed with only sufficient water to reduce dusting. This mixture is then forced through the delivery line to the nozzle by compressed air (figure 43). At the nozzle, sufficient water is added to the moving stream to meet the requirements of cement hydration. Figure 44 shows the nozzle and water ring of a dry mix shotcrete nozzle. For wet mix shotcrete, the cement, sand, and coarse aggregate are first conventionally mixed with water (figure 45), and the resulting concrete is then pumped to

the nozzle where compressed air propels the wet mixture onto the desired surface (figure 46). The two types of shotcrete produce mixes with different water contents and different application characteristics as a result of the distinctly different mixing processes. Dry mix shotcrete suffers high dust generation and rebound losses varying from about 15 percent to up to 50 percent. Wet mix shotcrete must contain enough water to permit pumping through the delivery line. Wet mix shotcrete, as a result, may experience significantly more cracking problems due to the excess water and drying shrinkage. Advances in the development of the high range water reducing admixtures, pumping aids, and concrete pumping equipment since about 1960 have greatly reduced these problems, and wet mix shotcrete is now being used more frequently in repair construction.

Shotcrete is a very versatile construction material that can be readily placed and successfully used for a variety of concrete repair applications. The necessity of form work can be eliminated in many repair applications by use of shotcrete. Shotcrete has been used to repair canal and spillway linings and walls, the faces of dams, tunnel linings, highway bridges and tunnels, deteriorating natural rock walls and earthen slopes, and to thicken and strengthen existing concrete structures. Provided the proper materials, equipment, and procedures are employed, such shotcrete repairs can be accomplished quickly and economically. This apparent ease of application should not cause one to believe that shotcrete repair is a simple procedure or one that can be haphazardly or improperly applied with impunity. The following two paragraphs contain a very descriptive warning of such practices:

"Regardless of the considerable advantages of the shotcrete process and its ability to provide finished work of the highest quality, a large amount of poor and sometimes unacceptable work has unfortunately occurred in the past, with the result that many design and construction professionals are hesitant to employ the process. As with all construction

Figure 43.—Dry mix shotcrete equipment being used in the Denver concrete laboratories.

Figure 44.—Dry mix shotcrete equipment showing the nozzle and water injection ring.

Figure 45.—Wet mix shotcrete equipment. The premixed shotcrete is delivered to the shotcrete pump by a transit truck.

Figure 46.—Wet mix shotcrete is propelled by compressed air.

methods, failure to employ proper procedures will result in inferior work. In the case of shotcrete the deficiencies can be severe, requiring complete removal and replacement.

Deficiencies in shotcrete applications usually fall into one of four categories: failure to bond to the receiving substrate, delamination at construction joints or faces of the application layers, incomplete filling of the material behind the reinforcing, and embedment of rebound or other unsatisfactory material." (Warner, 1995).

Each of the above-listed deficiencies has occurred on Reclamation repair projects. Perhaps more important with shotcrete than with any other standard concrete repair method, if highly qualified, well trained, and competent workmen cannot be employed, it is advisable to consider using some other repair procedure. The quality of shotcrete closely depends upon the skill and experience of one person, the nozzleman. Reclamation specifications require employment of only formally certified nozzlemen for shotcrete repairs. The on-the-job training necessary to develop the experience and skill needed to achieve such certification for Reclamation work should occur prior to the nozzleman's arrival at the job.

*(a) Preparation.*—Concrete to be repaired with shotcrete should be prepared in a manner identical to the preparation required for replacement concrete, section 29.(a). Experience indicates, however, that surface preparation for shotcrete repair is more critical than for replacement concrete. It is essential with shotcrete repairs that the shotcrete have a clean, sound concrete base for bond.

*(b) Materials.*—Cement used for shotcrete should meet the same requirements as cement used for replacement concrete, section 29.(b). If sulfate exposure conditions exist, type V portland cement should be specified. Normally, however, type I-II, low alkali cement is adequate.

Water, sand, and coarse aggregate used in shotcrete should also meet the requirements for replacement concrete, except that the maximum size coarse aggregate should not exceed 3/8 of an inch.

Additives for shotcrete should meet the requirements of ASTM designation C 494, Chemical Admixtures for Concrete. It is normally not possible to accomplish air entrainment with dry mix shotcrete. Lack of air entrainment may lead to dry mix shotcrete having lower than desired freeze-thaw resistance. Wet mix shotcrete should be proportioned to contain 6 to 8 percent entrained air.

It is sometimes desirable to use accelerating admixtures in shotcrete where rapid setting or rapid strength development is required. Calcium chloride accelerators have long been used, but there are now sufficient non-chloride containing accelerators in the marketplace to make the use of calcium chloride inadvisable. The use of calcium chloride accelerators is particularly unadvisable in shotcrete applications containing reinforcing steel or steel fibers.

Fiber reinforcement has been used in shotcrete since the early 1970s, and the M-47 specifications in appendix A contain specifications for steel fiber reinforcement. The American Concrete Institute has published a state of the art report on fiber reinforced shotcrete (American Concrete Institute, 1984), and this document should be consulted if the use of fiber reinforced shotcrete is being considered. It should be recognized that application of fiber reinforced shotcrete is more difficult and requires more experienced nozzlemen.

*(c) Application.*—The detailed discussion of shotcrete application techniques and technology is beyond the scope of this guide. The American Concrete Institute has published a recommended practice and a specification for materials, proportioning, and application of shotcrete (American Concrete

Institute, 1966; 1977). These documents should be studied before attempting shotcrete repairs.

*(d) Curing.*—Proper curing of shotcrete is essential if high strength properties, durability, and long service life are to be obtained. The M-47 specifications permit water curing or curing of shotcrete by application of curing compounds. It is important to begin curing by applying approved curing compounds or water spray before there has been evaporative water loss from the shotcrete, particularly during periods of high temperatures, low humidity, or high wind conditions. Improvements in bond strength will be obtained by continuing curing for periods of up to a month.

**29. Replacement Concrete.**—Concrete repairs made by bonding new concrete to repair areas without use of an epoxy bonding agent or mortar grout applied on the prepared surface should be made when the area exceeds 1 square foot and has a depth greater than 6 inches and when the repair will be of appreciable continuous area. Replacement concrete repairs should also be used for:

- Holes extending entirely through concrete sections

- Holes in which no reinforcement is encountered, or in which the depth extends 1 inch below or behind the backside of the reinforcing steel and which are greater in area than 1 square foot and deeper than 4 inches, except where epoxy- bonded concrete replacement is required or permitted as an alternative to concrete replacement

- Holes in reinforced concrete greater than one-half square foot and extending beyond reinforcement

Replacement concrete is the most common concrete repair material and will meet the needs of a majority of all concrete repairs. Replacement concrete repairs are made by bonding new concrete to the repair areas without the use of a bonding agent or portland

cement grout. The combination of a deep repair and good curing practices ensures adequate hydration water will remain at the bonding surface zone for at least 28 days, allowing the cement hydration process to develop good bond. Because the defective concrete is being replaced with high quality concrete very similar to the surrounding concrete, the repair is compatible in thermal expansion and in other physical and chemical properties with the old concrete. For this reason, in many cases, the best repair method is the use of replacement concrete. Only when an unusual increase in durability is needed, or when placing conditions or dimensions dictate otherwise, should other materials be considered.

*(a) Preparation.*—To obtain satisfactory results with the replacement concrete method, preparation should be as follows:

- Reinforcement bars should not be left partially embedded; concrete should be removed to provide a clearance of at least an inch around each bar exposed more than one-third its circumference.

- The perimeter of the hole at the face should be saw cut to a minimum depth of 1 inch. If the shape of the defect makes it advisable, the remainder of the concrete removal may be chipped below the vertical saw cut and continued until a horizontal surface is obtained. The top of the hole, if on a vertical wall, should be cut on a 1:3 upward slope from the back toward the face from which the concrete will be placed (see figure 14). This is essential to permit vibration of the concrete without leaving air pockets at the top of the repair. In some instances, where a hole extends through a wall or beam, it may be necessary to fill the hole from both sides; the slope of the top of the cut should be modified accordingly.

- The bottom and sides of the hole should be cut sharply and approximately square with the face of the wall. When the hole extends through the concrete section, spalling and feather edges must be avoided by having perimeter saw cuts from both faces. All interior corners should be rounded to a minimum radius of 1 inch.

- For repairs on surfaces subject to destructive water action and for other repairs on exposed surfaces, the outlines of areas to be repaired should be saw cut as directed to a depth of 1-1/2 inches before the defective concrete is excavated. The new concrete should be secured by keying methods. Figure 47 shows a vertical wall being prepared for replacement concrete repairs.

The construction and setting of forms are important steps in the procedure for satisfactory concrete replacement where the concrete must be placed from the side of the structure. Form details for walls are shown in figure 48. To obtain a tight and acceptable repair, the following requirements must be observed:

- Front forms for wall repairs more than 18 inches high should be constructed in horizontal sections so the concrete can be conveniently placed in lifts not more than 12 inches deep. The back form may be built in one piece. Sections to be set as concreting progresses should be fitted before placement is started.

- To exert pressure on the largest area of form sheathing, tie bolts should pass through wooden blocks fitted snugly between the walers and the sheathing.

- For irregularly shaped holes, chimneys may be required at more than one level; when beam connections are required, a chimney may be necessary on both sides of the wall or beam. For such construction, the chimney should extend the full width of the hole.

- Forms should be substantially constructed so that pressure may be applied to the chimney cap at the proper time.

- Forms must be mortar tight at all joints between adjacent sections, between the forms and concrete, and at tie bolt holes to prevent the loss of mortar when pressure is applied during the final stages of placement. Twisted or stranded caulking cotton, folded canvas strips, or similar material should be used as the forms are assembled.

Surfaces of old concrete to which new concrete is to be bonded must be clean, rough, and in a saturated surface dry condition. Extraneous material on the joint resulting from form construction must be removed prior to placement.

*(b) Materials.*—Concrete for repair should have the same water-cement ratio as used for similar new structures but should not exceed 0.47, by weight. Aggregate of as large a maximum size and slump as low as is consistent with proper placing and thorough vibration should be used to minimize water content and consequent shrinkage. The concrete should contain 3 to 5 percent entrained air. Where surface color is important, the cement should be carefully selected or blended with white cement to obtain the desired results. To minimize shrinkage, the concrete should be as cool as practicable when placed, preferably at about 70 °F or lower. Materials should, therefore, be kept in shaded areas during warm weather. Use of ice in mixing water may sometimes be necessary. Batching of materials should be by weight; but batch boxes, if of the exact size needed, may be used. Since batches for this class of work will be small, the uniformity of the materials is important and should receive proper attention.

Best repairs are obtained when the lowest practicable slump is used. This is about 3 inches for the first lift in an ordinary large form. Subsequent lifts can be drier, and the top few inches of concrete in the hole and that in the chimney should be placed at almost zero slump. It is usually best to mix enough

Figure 47.—Preparation of a wall for placement of replacement concrete repairs.

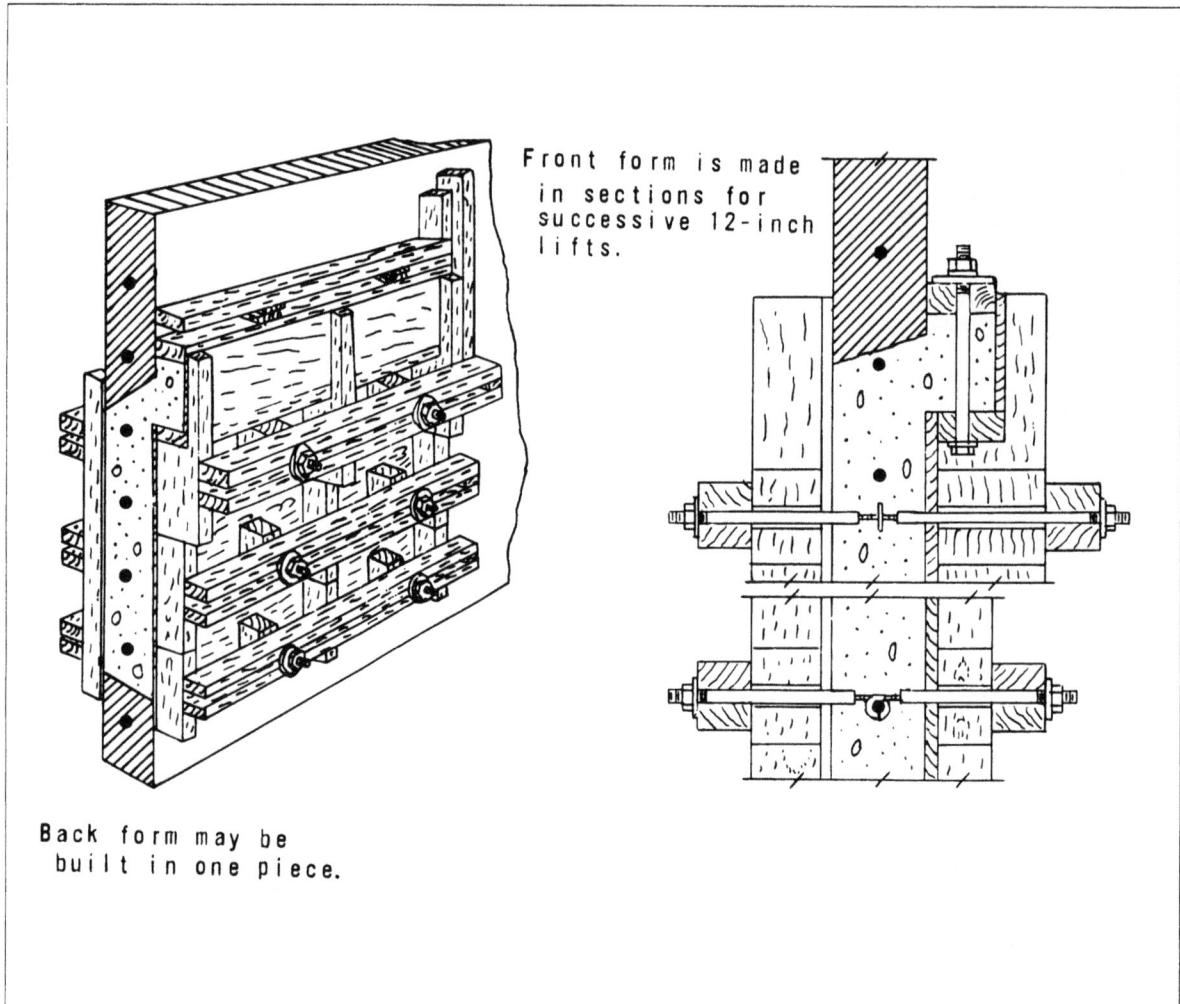

Figure 48.—Detail of forms for concrete replacement in walls.

concrete at the start for the entire hole. Thus, the concrete will be up to 1-1/2 hours old when the successive lifts are placed. Such premixed concrete, provided it can be vibrated satisfactorily, will have less settlement, less shrinkage, and greater strength than freshly mixed concrete.

Structural concrete placements should be started with an oversanded mix containing about a 3/4-inch-maximum size aggregate; a maximum water-cement ratio of 0.47, by weight; 6 percent total air, by volume of concrete; and having a maximum slump of 4 inches. This special mix should be placed several inches deep on the joint at the bottom

of the placement. A mortar layer should not be used on the construction joints.

*(c) Application.*—When placing concrete in lifts, placement should not be continuous; a minimum of 30 minutes should elapse between lifts. When chimneys are required at more than one level, the lower chimney should be filled and allowed to remain for 30 minutes between lifts. When chimneys are required on both faces of a wall or beam, concrete should be placed in only one of the chimneys until it flows to the other. Attempted placement in both chimneys will result in air entrapment and/or voids in the structure.

The quality of a repair depends not only on use of low-slump concrete, but also on the thoroughness of the vibration during and after depositing the concrete. There is little danger of overvibration. Immersion-type vibrators should be used if accessibility permits. If not, this type of vibrator can be used very effectively on the forms from the outside. Form vibrators can be used to good advantage on forms for large inaccessible repairs, especially on a one-piece back form, or attached to large metal fittings such as hinge-base castings. Immediately after the hole has been completely filled, pressure should be applied to the fill and the form vibrated. This operation should be repeated at 30-minute intervals until the concrete hardens and no longer responds to vibration. Pressure is applied by wedging or by tightening the bolts extending through the pressure cap (figure 48). In filling the top of the form, concrete to a depth of only 2 or 3 inches should be left in the chimney under the pressure cap. A greater depth tends to dissipate the pressure. After the hole has been filled and the pressure cap placed, the concrete should not be vibrated without a simultaneous application of pressure. To do so may produce a film of water at the top of the repair that will prevent bonding.

Addition of aluminum powder to concrete causes the latter to expand as described in section 182 of the *Concrete Manual* (Bureau of Reclamation, 1975). Under favorable conditions, this procedure has been successfully used to secure tight, well-bonded repairs in locations where the replacement material had to be introduced from the side. Forms similar to those shown in figure 48 should be used. Time should not be allowed for settlement between lifts. When the top lift and the chimney are filled, no pressure need be applied, but the pressure cap should be secured in position so expanding concrete will be confined to and completely fill the hole undergoing repair. There should be no subsequent revibration.

Concrete replacement in open-top forms, as used for reconstruction of the tops of walls, piers, parapets, and curbs, is a comparatively simple operation. Only such materials as will make concrete of proved durability should be used. The water-cement ratio should not exceed 0.47, by weight. For the best durability, the maximum size of aggregate should be the largest practicable and the percentage of sand the minimum practicable. No special features are required in the forms, but they should be mortar tight when vibrated and should give the new concrete a finish similar to the adjacent areas. The slump should be as low as practicable, and dosage of air entraining agent should be increased as necessary to secure the maximum permissible percentage of entrained air, despite the low slump. Top surfaces should be sloped to provide rapid drainage. Manipulation in finishing should be held to a minimum, and a wood-float finish is preferable to a steel-trowel finish. Edges and corners should be tooled or chamfered. Use of water for finishing is prohibited.

Forms for concrete replacement repairs usually may be removed the day after casting unless form removal would damage the green concrete, in which event stripping should be postponed another day or two. The projections left by the chimneys normally should be removed the second day. If the trimming is done earlier, the concrete tends to break back into the repair. These projections should always be removed by working up from the bottom because working down from the top tends to break concrete out of the repair. The rough area resulting from trimming should be filled and stoned to produce a surface comparable to that of surrounding areas. Plastering of these surfaces should never be permitted.

Some replacement concrete does not require forms. Replacement of damaged or deteriorated paving or canal lining slabs, wherein the full depth of the slab is replaced, involves procedures no different from those required for best results in original construction. Contact edges at the perimeter should be saw cut clean and square with the surface.

Special repair techniques are required for restoration of damaged or eroded surfaces of spillway or outlet works tunnel inverts and stilling basins. In addition to the usual forces of deterioration, such repairs often must withstand enormous dynamic and abrasive forces from fast-flowing water and sometimes from suspended solids. Silica fume concrete (section 37) is the repair material of choice for these types of repair. Whenever practicable, low slump silica fume concrete should be used. Slump of the concrete should not exceed 2 inches for slabs that are horizontal or nearly horizontal and 3 inches for all other concrete. (Note: This is 1 inch less slump than that permitted in the M-47 specifications for conventional applications of silica fume concrete.) The net water-cementitious ratio (exclusive of water absorbed by the aggregates) should not exceed 0.35 by weight. An air-entraining agent and a high range water reducing admixture should be used. Set-retarding admixtures should be used only when the interval between mixing and placing is quite long. (These recommendations are repeated in section 37.)

If, however, needed repairs are too small for the replacement concrete method (including silica fume concrete), they should be made using the dry pack procedure (section 26), the epoxy-bonded epoxy mortar method, (section 30) or the epoxy-bonded replacement concrete method (section 31).

*(d) Curing and Protection.*—The importance of curing replacement concrete repairs cannot be overemphasized. Complete failure of repairs has been attributed to inadequate or improper curing. There is no known condition short of flooding the repaired structure (which in itself is an excellent curing method) that does not require curing of cementitious repairs. Because of the relatively small volume of most repairs and the tendency of old concrete to absorb moisture from new material, water curing is a highly desirable procedure, at least during the first 24 hours. When forms are used for repair, they can be removed and then reset to hold a few layers of wet burlap in contact with new concrete. One of the best methods of water curing is a soil-soaker hose laid beneath a plastic membrane covering the repair area.

When curing compound is used, the best curing combination is an initial water-curing period of 7 days (never less than 24 hours) followed, while the surface is still damp, by a uniform coat of the compound. It is always essential that repairs, even dry packed cone bolt holes, receive some water curing and be thoroughly damp before the curing compound is applied. If nothing better can be devised for the initial water curing of the dry pack in cone bolt holes and similar repairs, a reliable workman should be detailed to make the rounds with water and a large brush or a spraying device to keep the repaired surfaces wet for 24 hours prior to application of a curing compound. White curing compound may be used only where its color does not create objectionable contrast in appearance.

**30. Epoxy-Bonded Epoxy Mortar.**— Epoxy-bonded epoxy mortar should be used where the depth of repair is less than 1-1/2 inches and the exposure conditions are such that *relatively constant temperatures* can be expected. Epoxy mortars have thermal coefficients of expansion that may be significantly different from conventional concrete. If such mortars are used under conditions of wide and frequent temperature fluctuations, they will cause failure just below the bond surface in the base concrete. *For this reason, current Reclamation practice precludes the use of epoxy mortars under conditions of frequent or large temperature fluctuations.*

The application of epoxy mortar to repair areas of concrete deterioration caused by corroding reinforcing steel is also not recommended. The epoxy bond coat and epoxy mortar create zones of electrical potential that are different from the electrical potential in the surrounding concrete. This difference in potential can result in the formation of a galvanic corrosion cell with accelerated corrosion at the repair perimeters.

Epoxy mortar is properly used to make thin repairs (1/2-inch to 1-1/2-inch thickness) to concrete under relatively constant temperature exposure conditions. Such applications could include tunnel linings, indoor or interior concrete, the underside of concrete structures such as bridge decks, continuously inundated concrete such as stilling basin floors, canal linings below water line, or concrete pipe. Applications to concrete exposed to the daily temperature fluctuations caused by exposure to direct sunlight *are not appropriate for epoxy mortar repair.*

Properly applied epoxy mortar repairs have a long history of successful performance on Reclamation concrete when used under appropriate conditions. A 1991 inspection of the epoxy mortar repairs made at Yellowtail Dam in 1968 showed that less than 2 percent of the repairs had suffered failure in over 20 years of service. This is considered out-standing performance for a repair material.

*(a) Preparation.*—Concrete to be repaired with epoxy mortar should be prepared in accordance with section 8. Prior to application of the epoxy mortar, the concrete should be heated in sufficient depth, when necessary, so that the surface temperature (as measured by a surface temperature gage) does not drop below 40 °F during the first 4 hours after placement of an epoxy bond coat. This may require several hours of preheating with radiant heaters or other approved means (figures 49 and 50). If existing conditions prohibit meeting these temperature require-ments, suitable modifications should be adopted upon the approval of the inspector or other responsible official. The concrete temperature during preheating should never exceed 200 °F, and the final surface tempera-ture at the time of placing epoxy materials should never be greater than 100 °F.

*(b) Materials.*—Epoxy resins used to prepare epoxy mortar for use in concrete repair should be two-component, 100-percent solids type meeting the requirements of specification ASTM C-881 for type III, grade 2, class B or C. Class B epoxy is used between 40 and 60 °F. Class C epoxy is used above 60 °F up to the highest temperature defined by the epoxy manufacturer.

The sand used in epoxy mortar must be clean, dry, well graded, and composed of sound particles. For most applications, sand passing a No. 16 screen and conforming to the following limits should be used:

| Screen number | Individual percent, by mass, retained on screen |
|---|---|
| 30 | 26 to 36 |
| 50 | 18 to 28 |
| 100 | 11 to 21 |
| Pan | 25 to 35 |

Range shown is applicable when 60 to 100 percent of pan is retained on No. 200 screen. When 41 to 100 percent of pan passes the No. 200 screen, the percent pan should be within the range of 10 to 20 percent, and the individual percentages retained on the Nos. 30, 50, and 100 screens should be adjusted accordingly.

Sand processed for use in concrete rarely contains the required quantity of pan size sand. As a result, problems often arise in obtaining additional pan size material to supplement sand available on the jobsite. A source of silica pan size material may be obtained by contacting the Materials Engineering and Research Laboratory, Code D-8180, Bureau of Reclamation, Denver Federal Center, Denver, Colorado 80225. A sand graded as shown above and properly mixed with an epoxy meeting ASTM C-881 specifications will provide a dense, high strength, workable epoxy mortar.

The sand should be maintained in a dry area at not less than 70 °F for 24 hours immediately prior to the time of use. Filler materials other than sand, such as portland cement, can be used. However, for general applications, a natural sand is recommended.

It is also acceptable to obtain and use brand name prepackaged epoxy mortar repair systems that contain resin and sand, *provided that the resin systems meet the ASTM C-881 specifications previously listed.* Such mortar

Figure 49.—A gas-fired forced air heater is being used to heat concrete
prior to application of epoxy mortar.

Figure 50.—An enclosure has been constructed over an area to be repaired with
epoxy mortar to keep the concrete warm.

systems are manufactured specifically for concrete repair and must be used in exact accordance with the manufacturer's recommendations.

On critical repair jobs such as areas of high velocity flow or on repairs requiring a considerable quantity of materials, the contractor should be required to submit samples of epoxy resin and graded sand to the Materials Engineering and Research Laboratory, Denver, for use in mix design determinations. The samples should consist of 1 gallon total quantity of epoxy components and a minimum of 50 pounds of graded sand. Samples should be submitted at least 30 days prior to use in the work and be labeled or otherwise identified with the specifications number under which the material is to be used.

*(c) Mixing.*—Preparation of epoxy mortar involves premixing proper quantities of epoxy resin and hardener and then mixing the resin system with sand to make the epoxy mortar.

The epoxy resin used for mortar preparation is a two-component (part A and part B) material which requires accurate combination of components and mixing prior to use. Once mixed, the material has a limited pot life and must be used immediately. (Pot life refers to the period of time elapsing between mixing of ingredients and their stiffening to the point where satisfactory use cannot be achieved.) The repair resin should be prepared by adding the required quantity of hardener (normally, part B) to the resin (normally part A) in proportions recommended by the manufacturer, followed by thorough mixing. Since the pot life of the mixture depends on the temperature (longer at low temperature, much shorter at high temperature), the quantity to be mixed at one time should be that quantity that can be applied within approximately 30 minutes. The addition of nonreactive thinners or diluents to the resin mixture is not permitted since it weakens the epoxy.

The epoxy mortar is composed of sand and epoxy resin suitably blended to provide a stiff, workable mix. Mix proportions should be established, batched, and reported on a weight basis, although the dry sand and mixed epoxy may be batched by volume using suitable measuring containers that have been calibrated on a weight basis. Epoxy meeting ASTM specification C-881 will require approximately 5-1/2 to 6 parts of graded sand to 1 part epoxy, by weight. This is equivalent to a ratio of approximately 4 to 4-1/2 parts sand to 1 part epoxy, by volume. If equivalent volume proportions are being used, care must be taken to prevent confusing them with weight proportions. It will be necessary to adjust the mix proportions for the particular epoxy and sand being used. The epoxy mortar should be thoroughly mixed with a slow-speed mechanical device. The mortar should be mixed in small size batches so that each batch can be completely mixed and placed within approximately 30 minutes. Figure 51 shows a simple bucket mixer that is adequate to mix epoxy mortar for small repairs.

*(d) Application.*—Application of epoxy mortar repairs first requires application of a resin bond coat followed by application and finishing of the epoxy mortar. Surfaces of existing concrete to which epoxy mortar is to be bonded should be prepared as discussed in section 8. Steel to be embedded in epoxy mortar should be prepared, cleaned, and dried in the same manner as the concrete being repaired. The exposed steel should be completely coated with epoxy bonding agent when the agent is applied to the surfaces of the repair area.

A resin bond coat consisting of the same type epoxy resin used to mix the epoxy mortar is applied to the prepared concrete surface immediately before placing the epoxy mortar. After the bond coat resin is mixed, it must be uniformly applied to the prepared, dry, existing concrete at a coverage of not more than 80 square feet per gallon, depending on surface conditions. The area of coverage per gallon of resin depends on the roughness of the surface to be covered and may be considerably less than the maximum specified. The epoxy bonding agent may be applied by any convenient, safe method such as

Figure 51.—A bucket mixer can be used to mix epoxy mortar for small repair areas.

squeegee, brushes, or rollers which will yield an effective coverage. Spraying of the material is permitted if an efficient airless spray is used and if the concrete surfaces to receive the agent are at a temperature of 70 °F or somewhat warmer. Before approving spraying, it should be demonstrated that spraying will provide an adequate job with minimum overspray. If spray application is used, the operator must wear a compressed air-fed hood, and no other personnel should be closer than 100 feet if downwind of the operator.

During application of the epoxy bond coat, care must be exercised to confine the material to the area being bonded and to avoid contamination of adjacent surfaces. However, the bond coat should extend slightly beyond the edges of the repair area.

The applied epoxy bonding resin must be in a fluid condition when the epoxy mortar is placed. If the resin cures beyond this fluid state but is still tacky, a second bond coat should be applied over the first coat. If any bond coat has cured beyond the tacky state, it must be completely removed by sandblasting, the concrete properly cleaned, and a new bond coat applied.

Special care must be taken to prevent the bond coat from being spread over concrete surfaces not properly cleaned and prepared.

Appropriate solvents may be used to clean tools and spray guns, but in no case should the solvents be incorporated in any bonding agent. All tools must be completely dried before reuse.

The prepared epoxy mortar should be tamped, flattened, and smoothed into place (figure 52) in all areas while the bonding resin is still in a fluid condition, except that on steep slopes, the bond coat can be allowed to stiffen to a very tacky condition to assist in holding the mortar in place. Special care must be taken to thoroughly compact the epoxy mortar against the bond coat. The mortar should be worked to grade and given a steel trowel finish (figure 53). Special care must be taken at the edges of the area being repaired to assure

complete filling and leveling and to prevent the mortar from being spread over surfaces not having the epoxy bond coat application. Steel troweling should best suit prevailing conditions; in general, it should be performed by applying slow, even strokes. Trowels may be heated to facilitate the finishing, but the use of thinner, diluents, water, or other lubricants on placing or finishing tools is not permitted. After leveling the epoxy mortar to the finished grade where precision surfaces are required on sloping, vertical, or overhead surfaces, the mortar should be covered with plywood panels smoothly lined with polyethylene sheeting and weighted with sandbags or otherwise braced by suitable means until the possibility of slumping has passed. When polyethylene sheeting is used, no attempt should be made to remove it from the epoxy mortar repair before final hardening.

Surfaces of all epoxy mortar repairs should be finished to the plane of surfaces adjoining the repair areas. The final finished surfaces should have the same smoothness and texture of surfaces adjoining the repair areas.

*(e) Curing.*—Epoxy mortar repairs should be cured immediately after completion at not less than the temperature range prescribed by the class of the epoxy until the mortar is hard. Postcuring, if required by the specifications, can then be initiated at elevated temperatures by heating in depth the epoxy mortar and the concrete beneath the repair. Postcuring should continue for a minimum of 4 hours at a surface temperature generally not less than 90 °F nor more than 110 °F. The heat could be supplied by use of portable propane-fired heaters, infrared lamp heaters, or other approved sources positioned to attain the required surface temperatures (figure 54).

In no case should epoxy-bonded epoxy mortar be subjected to moisture until after the specified postcuring has been completed.

Epoxy mortars generally produce patches that are darker than the surrounding concrete. Some available epoxies produce a gray-colored mortar resembling concrete. However, these materials will rarely produce

Figure 52.—Epoxy mortar is consolidated and compacted by hand tamping.

Figure 53.—Applying the steel trowel finish required by epoxy mortar repairs.

Figure 54.—Postcuring heating enclosure installed over an epoxy mortar repair area.

an exact color match. Grinding hardened epoxy mortar may lighten its color to about that of the surfaces adjoining the repair areas. Epoxy mortars can be colored by the addition of such materials as iron oxide red, chromium oxide green, lampblack and titanium dioxide white for gray, and ocher yellow; although Reclamation rarely uses any materials to color the epoxy other than the sand for the mortar. Use of white silica sand in the mortar will produce a white-looking patch; most natural riverborne sands will produce darker colored mortars. Whenever epoxy mortar repair materials must be colored to match adjacent concrete, laboratory mixes should be made to ascertain the proper quantities of coloring constituents.

*(f) Safety.*—All personnel must be carefully instructed to take every precaution in preventing epoxy resins and their components from contacting the skin and in preventing the breathing of epoxy fumes or vapors. Protective clothing must be worn, including gloves and goggles, and protective creams for other exposed skin areas should be provided when handling epoxies, as severe allergic reactions and possible permanent health damage can result when these materials are allowed to

contact and remain upon the skin. Any deposits acquired through accidental contact of these materials with unprotected skin must be removed immediately by washing with soap and water, never with solvents. Solvents, such as toluene and xylene, may be used only for cleaning epoxy from tools and equipment. Care must also be exercised to avoid contact of cleaning solvents with the skin and to provide adequate ventilation for mixing, placing, and cleanup operations. All safety equipment used must conform to the requirements of the Occupational Safety and Health Standards of the Occupational Safety and Health Administration.

**31. Epoxy-Bonded Replacement Concrete.**— Epoxy-bonded concrete is used for repairs to concrete that are between 1.5 and 6 inches thick. Shallow replacement concrete repairs, less than 6 inches thick, are subject to poor curing conditions as a result of moisture loss to evaporation and to capillary absorption by the old base concrete. Such repairs seldom develop acceptable bond strength to the old concrete. The epoxy bonding resin is used to ensure a strong, durable bond between the old concrete and the replacement concrete.

As with epoxy-bonded epoxy mortar, care should be exercised if epoxy-bonded concrete is to be used to repair shallow deterioration resulting from corroding reinforcement. The epoxy bond coat may create electrical potentials sufficiently different from potentials in the surrounding concrete to result in accelerated corrosion at repair perimeters.

*(a) Preparation.*—Concrete to be repaired with epoxy-bonded concrete must be prepared as described in section 8.

*(b) Materials.*—The materials used in epoxy-bonded concrete repairs consist of conventional portland cement concrete and epoxy resin bonding agent.

The concrete used for epoxy-bonded repairs is the same as that used for replacement concrete repairs (section 29) except that the slump of the concrete when placed should not exceed 1-1/2 inches.

A number of proprietary epoxy formulations prepared for bonding new concrete to old concrete are now available. Many of these materials are excellent high quality products and can be used with reasonable certainty as to the results. However, some of the resins available are unsuitable or untested for such repair applications, and care should be taken to use only the epoxy bonding resins meeting the requirements of specification ASTM C-881 for a type II, grade 2, class B or C epoxy system. Class B epoxy should be used when the temperatures are above 40 °F but less than 60 °F. Class C epoxy should be used when concrete temperatures are from 60 °F up to the maximum temperature recommended by the epoxy manufacturer.

The epoxy resin used for epoxy-bonded concrete is a two component, 100-percent solids resin system requiring accurate proportioning and thorough mixing prior to use. The procedures described in section 30 should be followed during preparation and application of the resin. Conventional concrete mixing procedures as described in section 29 should be followed to mix the concrete.

*(c) Application.*—Use of epoxy-bonded concrete in repairs requiring forming, such as on steeply sloped or vertical surfaces, can be permitted only when sufficient time has been allowed to place concrete against the epoxy bonding resin while it is still fluid. If the resin cures before placement of the concrete, no bond will develop between the old and new concrete. It is a good idea to practice install such forms at least once before actually applying the epoxy bond coat (figure 55).

Immediately after application of the epoxy resin bonding agent and while the epoxy is still fluid, unformed epoxy-bonded concrete should be spread evenly to a level slightly above grade and compacted thoroughly by vibrating or tamping (figure 56). Tampers should be sufficiently heavy for thorough compaction. After being compacted and screeded, the concrete should be given a wood-float or steel-trowel finish as required. Water, cement, or a mixture of dry cement and sand should never be sprinkled on the surface. Troweling, if required, should be performed at the proper time and with heavy pressure to produce a smooth, dense finish free of defects and blemishes. As the concrete continues to harden, the surface should be given additional trowelings.

The final troweling should be performed after the surface has hardened so that no cement paste will adhere to the edge of the trowel, but excessive troweling cannot be permitted.

*(d) Curing.*—Even though an epoxy bond coat is used, it still remains essential to properly cure epoxy-bonded concrete. As soon as the epoxy-bonded concrete has hardened sufficiently to prevent damage, the surface should be cured by spraying lightly with water and then covering with sheet polyethylene or by coating with an approved curing compound. Curing compound should be used whenever there is any possibility that freezing temperatures will prevail during the curing period. Sheet polyethylene must be an airtight, nonstaining, waterproof covering that will effectively prevent evaporation. Edges of the polyethylene should be lapped and sealed. The waterproof covering should be left in

Figure 55.—If forms are required for epoxy-bonded concrete repairs, they should be installed at least once prior to application of the epoxy bond coat to ensure that they fit as planned and that they can be installed and filled before the bond coat hardens.

Figure 56.—The placement techniques for epoxy-bonded concrete are essentially the same as for conventional concrete.

place for at least 2 weeks. If a waterproof covering is used and the concrete is subjected to any usage during the curing period that might rupture or otherwise damage the covering, the covering must be protected by a suitable layer of clean, wet sand or other cushioning material that will not stain concrete. Application of curing compound must be in accordance with appropriate standard procedures as contained in the *Concrete Manual* (Bureau of Reclamation, 1975).

*(e) Safety.*—All personnel must be carefully instructed to take every precaution in preventing epoxy resins and their components from contacting the skin and in preventing the breathing of epoxy fumes or vapors. Protective clothing must be worn, including gloves and goggles, and protective creams for other exposed skin areas should be provided when handling epoxies, as severe allergic reactions and possible permanent health damage can result when these materials are allowed to contact and remain upon the skin. Any deposits acquired through accidental contact of these materials with unprotected

skin must be removed immediately by washing with soap and water, never with solvents. Solvents, such as toluene and xylene, may be used only for cleaning epoxy from tools and equipment. Care must also be exercised to avoid contact of cleaning solvents with the skin and to provide adequate ventilation for mixing, placing, and cleanup operations. All safety equipment used must conform to the requirements of the Occupational Safety and Health Standards of the Occupational Safety and Health Administration.

**32. Polymer Concrete.**—Polymer concrete (PC) is a concrete system composed of a polymeric resin binder and fine and coarse aggregate. Water is not used to mix polymer concrete. Instead, the liquid resin, known as a monomer, is caused to cure or harden by a chemical reaction known as polymerization. During polymerization, the monomer molecules are chemically linked and cross linked to form a hard, glassy plastic known as a polymer. The polymers used in PC are formulated to provide the special properties needed for high performance repair materials.

These systems can be cured very quickly and are most useful in performing repairs to structures that must be immediately returned to service. As an example, PC is commonly used to repair potholes in concrete highway bridge decks, thereby eliminating the necessity of long and costly road closures or detours. It is also useful for repairs to structures, such as tunnel linings, that can be maintained in a dry condition for only short periods of time and for cold weather repairs down to temperatures as low as 15 °F. PC repairs can be accomplished in thicknesses varying from about 1/2 inch to several feet if appropriate precautions are taken.

PC develops strength and durability properties very quickly due to its rapid polymerization characteristics and is useful where rapid repairs to concrete are required. PC can be mixed, placed, polymerized, and put into service in only a matter of hours. PC also develops enhanced durability properties. This feature makes it useful as protective overlays on conventional concrete exposed to corrosive or severe environments. Since PC does not contain mix water, it can be used at much lower temperatures (down to 15 °F) than portland cement concrete.

Most polymer concretes experience some volumetric shrinkage during polymerization and also have problems associated with the coefficient of thermal expansion similar to those experienced with epoxy mortars. These problems with PC, though generally less severe than similar problems with epoxy mortar, can limit the materials use on concrete exposed to wide temperature variations. Potential users of PC should be aware of these problems. The Materials Engineering and Research Laboratory at Denver, D-8180, can provide guidance and recommendations for the application of these very useful materials.

*(a) Preparation.*—Concrete preparation for PC repairs should be in accordance with section 8. Although some manufacturers indicate that PC may be used for feather edge repairs, Reclamation experience is otherwise. Saw cut repair perimeters are required if high quality repairs are to be achieved. Special care should

be taken to ensure the base concrete is dry prior to application of the repair material. Once the PC is in place, this requirement can be relaxed. However, flowing water may remove the fresh polymer concrete if allowed on the concrete prior to development of initial set.

*(b) Materials.*—A number of manufacturers have developed prepackaged PC systems. Most of the systems consist of acrylic or vinyl ester monomers, appropriate polymerization initiators and catalysts, and fine aggregate and fillers. It is common for the user to extend the polymer concrete, particularly when used to fill depressions deeper than 1 inch, by supplying and adding coarse aggregate to the prepackaged PC system during mixing. The prepackaged polymer concretes also contain monomer bond coat systems that must be mixed and applied to the base concrete prior to application of the polymer concrete mixture.

*(c) Application.*—Each manufacturer of prepackaged PC provides detailed instructions for proportioning, mixing, and applying its product. These instructions must be closely followed to obtain a satisfactory repair.

A bond coat monomer system is mixed and applied to the prepared concrete prior to application of the PC. Care must be taken to proportion and mix the bond coat components properly. Some manufacturers specify that the bond coat be applied and cured prior to placing the polymer concrete. Others specify that the polymer concrete be applied while the bond coat is still in the liquid state. It is important to follow the procedures recommended by the manufacturer of the product actually being used.

Polymer concrete can be mixed in paddle-type, rotary drum-type, or other types of power equipment suitable for mixing conventional concrete. Very small quantities of PC can even be mixed in the original shipping containers. Three minutes of mixing time should be adequate for all the prepackaged systems. It is common practice to first add the dry powder and aggregate components to the mixer and then add the liquid resin. With some systems, the liquid component may require a separate premixing step to combine the monomer with the catalyst or initiators needed for polymerization. Other manufacturers include the initiators in the powder-fine aggregate component, thereby eliminating the premixing requirement.

The mixed PC is placed just like conventional concrete, using the same tools and procedures (figure 57). Most PC mixtures will be almost self-leveling and require only a minimum finishing operation. Light mechanical vibration should be provided to consolidate placements thicker than 2 to 3 inches (figure 58). Once the PC has been placed and consolidated, it should be screeded to proper grade and quickly finished with a wood or steel trowel. The top surface of the repair forms a "skin" soon after being placed. If repeated toweling is attempted, this skin will tear and cause an unsightly surface.

Polymer concrete develops a strong bond to most materials. If forms are used, they must be leakproof and provided with some method of bond breaker or release agent. Wrapping the forms with polyethylene film has proven a very effective method of preventing bond between the form and the PC.

**(d) Curing.**—Polymer concretes polymerize and harden very quickly under most ambient conditions and will develop nearly full strength within a 1- to 2-hour period. During this time, the fresh concrete must be protected from water and not disturbed (figure 59). If temperatures are lower than about 40 °F, the polymerization reaction will occur at a slower rate unless increased concentrations of initiator are used or the repair is heated to 70° to 80 °F. Conversely, polymerization can occur too quickly, with insufficient finishing time, if the ambient temperature exceeds 90° to 100 °F. A reduction of initiator concentration can reduce this problem. Alternately, the repairs can be made during the cooler parts of the day or at night.

**(e) Safety.**—All workers, supervisors, and inspectors involved with the project must be made aware of the procedures required for safe use of PC. The manufacturers of PC provide recommendations for safe storage, handling, and use. These recommendations must be known and followed by users of the materials. The following *minimum* safety requirements must be followed on Reclamation projects:

*Storage.*—Polymer concrete monomer and initiators are heat sensitive and flammable. These materials should be stored away from the direct rays of sunlight, in the original shipping containers, in well-ventilated areas away from sources of ignition. The storage temperature should not exceed 80 °F. Storage periods should not exceed manufacturer's recommendations.

*Mixing and Handling.*—Smoking, flame, or other sources of ignition must not be permitted during mixing and application. Electrical equipment in contact with polymer concrete should be grounded for safe discharge of static electricity. Type B or type ABC fire extinguishers must be provided at the storage, mixing, and application locations.

*Personal Protective Equipment.*—Workers using polymer concrete must be provided rubber boots and required to use disposable protective clothing. Splash-type safety goggles and impervious gloves must be provided to workers using polymer concrete, and the workers must be required to wear these items. In some instances where ventilation is poor or inadequate, workers may be required to wear organic vapor respirators. The mixing and application site should be provided with portable eyewash equipment capable of sustaining a 15-minute stream of clean, room temperature water.

**33. Thin Polymer Concrete Overlay.**—The thin PC overlay is a hard, glassy concrete coating, 25 to 50 mils thick, consisting of a vinyl ester resin system, silica flour filler,

Figure 57.—Placing polymer concrete in a repair area. Sandbags and polyethylene sheeting were used to prevent water from entering the repair area.

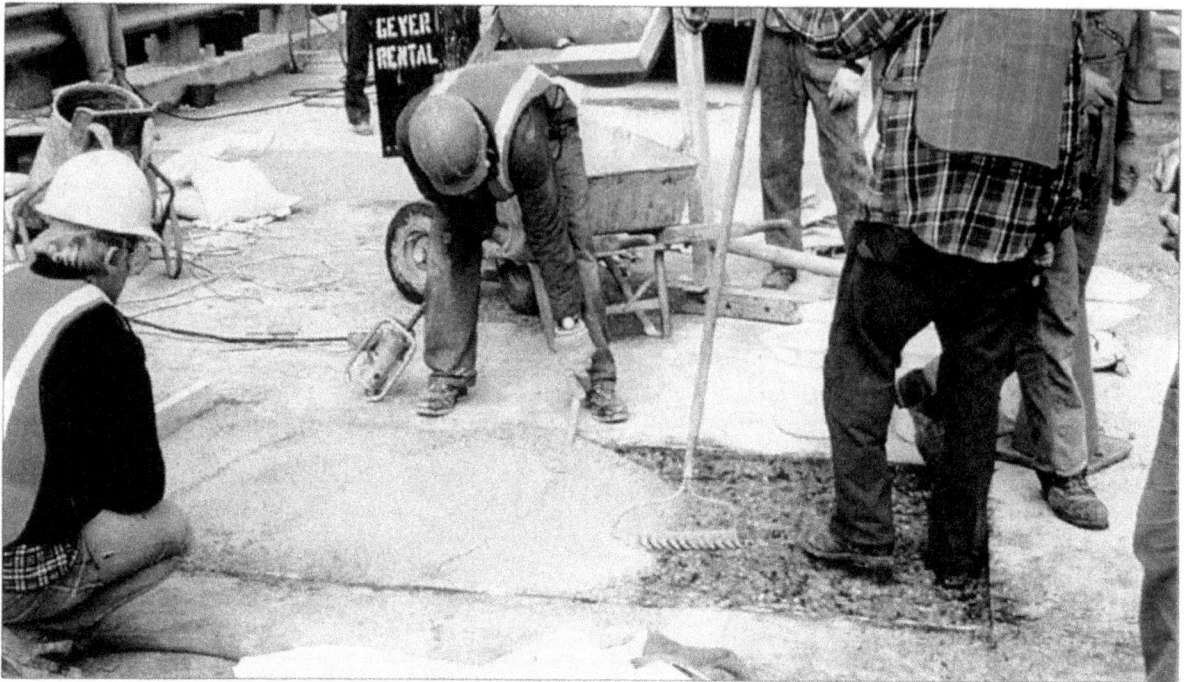

Figure 58.—Small stinger vibrators can be used to consolidate shallow depths of polymer concrete.

Figure 59.—Polymer concrete must be protected from water and not disturbed during the 1- to 2-hour curing period.  No other curing procedures are required unless ambient temperatures are very low.

and appropriate coloring pigments. This membrane-forming overlay partially penetrates the immediate top surface of the concrete and provides very good protection to concrete exposed to adverse chemical or weathering conditions. It can also provide cosmetic treatment to concrete exposed to public view. The normal three-coat application of this material (one primer coat plus two filler coats) should result in a total overlay thickness of about 50 mils.

The overlay is applied to protect the concrete from water penetration and resulting freeze-thaw damage; to protect the concrete from chemical corrosive elements such as acids, chlorides, or sulfates; and/or to improve the cosmetic appearance of the concrete. The overlay provides complete opaque coverage of the concrete surface and flows into and seals narrow cracks in the surface. The thin PC overlay is a standard repair material specified in the *Standard Specifications for the Repair of Concrete, M-47* of appendix A. There is currently no known commercial manufacturer of the thin polymer concrete overlay. Contractors or users of the overlay system can have the material prepared by custom resin blenders or can blend and mix the material themselves using the formulas listed below.

**(a) Preparation.**—Concrete to receive the thin polymer concrete overlay must be cleaned and prepared using light hydroblasting or wet sandblasting in accordance with the requirements of section 8. The prepared concrete surfaces must then be maintained in a clean, dry condition until the placement of the thin overlay is completed. The dryness of the concrete can be checked by taping a clear polyethylene sheet onto the surface of the concrete in a sunlight exposure. If no moisture collects under the polyethylene sheet after an hour or two of sunlight, the surface is sufficiently dry.

**(b) Materials.**—The thin overlay is prepared with vinyl ester resin, polymerization initiator and promoter, silicon flour filler, and pigments.

*Vinyl Ester.*—The vinyl ester resin is Dow Derakane 8084, manufactured by Dow Chemical Co., 2800 Mitchell Drive, Walnut Creek, California 94596.

*Initiator.*—The initiator is cumene hydroperoxide-78 percent, manufactured by Lucidol Division, Penwalt Corp., 1740 Military Road, Buffalo, New York 14240, or equal.

*Promoter.*—The promoter is cobalt napthenate-6 percent, available from fiber-glass materials suppliers.

*Filler.*—The filler is ground silica, minus 45 micrometers (No. 355) sieve size, manufactured by Ottawa Silica Co., Ottawa, Illinois; or Silco Seal 395 Ground Silica, manufactured by VWR Scientific, PO Box 3200, San Francisco, California 94119.

*Pigment.*—Two pigments are required to obtain a concrete gray color.

Titanium dioxide powder, manufactured by VWR Scientific

Carbon lamp black or bone black powder (do not use activated carbon)

**Mixing Proportions.**—

| Primer | 5.0 | gallons vinyl ester resin |
|---|---|---|
| | 0.60 | pounds initiator |
| | 0.25 | pounds promoter |
| Pigmented topcoats | 5.0 | gallons vinyl ester resin |
| | 1.35 | pounds initiator |
| | 0.27 | pounds promoter |
| | 40.0 | pounds filler |
| | 4.00 | pounds titanium dioxide pigment |
| | 0.02 | pounds carbon black pigment |

*Mixing Sequence.*—The vinyl ester resin, filler, and pigments should be premixed and set aside for several hours to wet out before use. Immediately prior to application, the initiator is added and thoroughly mixed with the resin. Then, the promoter is added to the resin and thoroughly mixed. *Never directly mix initiator and promoter, or an extremely violent and explosive reaction will occur.*

*(c) Application.*—The thin polymer concrete overlay consists of one primer coat and one or two sealant coats applied at a coverage rate of 1.3 to 2.0 gallons of material per 100 square feet of surface per coat, depending on surface texture of the concrete. The material may be applied with brooms, brushes, or paint rollers (figure 60).

The primer must uniformly and completely cover the surface and should be scrubbed into the surface of the concrete, eliminating discontinuities and puddles. The pigmented topcoat(s) must be applied to the primed surface not less than 4 hours nor more than 24 hours after application of the primer or of a succeeding to coat. It is not desired that the primer or topcoat cure to a hard final finish before application of the succeeding topcoat. A somewhat tacky finish is more desirable. However, full bond between coats will be obtained by following the above-listed timeframe for application.

Caution must be exercised to prevent application of primer or topcoats not containing initiator, promoter, or both. A simple technique of reducing the possibility of this is to place containers of preweighed resin, initiator, and promoter at appropriate locations along the application route before the application begins. After the containers are preplaced, one workman can quickly check that all needed components are at each location. Then, as applicators need a resupply of the primer or topcoating system, they need only to mix all the containers at the next location.

Application of the thin PC overlay system proceeds quite quickly. By proper pre-planning, two men were able to prime and topcoat the power house roof shown in figure 61 in 2 days.

*(d) Curing and Protection.*—The coated surfaces must be protected until the resin has completely cured to a hard finish. Such condition will normally be obtained within about 30 hours of application of the final topcoat. Low ambient temperatures and/or high relative humidity may lengthen the hardening process.

*(e) Safety.*—All workers, supervisors, and inspectors involved with the project must be made aware of the procedures required for safe use of the thin polymer concrete coating. The manufacturers of vinyl ester resin and the initiators and promoters provide recommendations for safe storage, handling, and use. These recommendations must be known and followed by users of the materials. The following *minimum* safety requirements must be followed on Reclamation projects:

*Storage.*—Vinyl ester resin and initiators are heat sensitive and flammable. These materials should be stored away from direct sunlight, in the original shipping containers, in well-ventilated areas away from sources of ignition. The storage temperature should not exceed 80 °F. Storage periods should not exceed manufacturer's recommendations.

*Mixing and Handling.*—Smoking, flame, or other sources of ignition must not be permitted during mixing and application. Electrical equipment in contact with polymer concrete should be grounded for safe discharge of static electricity. Type B or type ABC fire extinguishers must be provided at the storage, mixing, and application locations. *Never directly mix initiator and promoter, or an extremely violent and explosive reaction will occur.*

*Personal Protective Equipment.*—Workers using thin polymer concrete overlays must be provided rubber boots and required to use disposable protective clothing. Splash-type safety goggles and impervious gloves must be provided to workers using polymer concrete, and the workers must be required to wear these items. In some instances where ventilation is poor or inadequate, workers may be required to wear organic vapor respirators. The mixing and application site should be provided with portable eyewash

Figure 60.—The thin PC overlay system may be applied with push brooms, squeegees, or heavy industrial grade paint rollers.

Figure 61.—The thin PC overlay system can be applied very quickly.  Two workmen completed application to this powerplant roof in 2 days.

equipment capable of sustaining a 15-minute stream of clean, room temperature water.

**34.  Resin Injection.**—Resin injection is used to repair concrete that is cracked or delaminated and to seal cracks in concrete to water leakage.  Two basic types of resin and injection techniques are used to repair Reclamation concrete.

*(a)  Epoxy Resins.*—Epoxy resins cure to form solids with high strength and relatively high moduli of elasticity.  These materials bond readily to concrete and are capable, when properly applied, of restoring the original structural strength to cracked concrete.  The high modulus of elasticity causes epoxy resin systems to be unsuitable for rebonding cracked concrete that will undergo subsequent movement.  Epoxy resin has been used to seal cracks in concrete to waterflow.  The epoxies, however, do not cure very quickly, particularly at low temperatures, and using them to stop large flows of water may not be practical.  Cracks to be injected with epoxy resins should be between 0.005 inch and 0.25 inch in width.  It is difficult or impossible to inject resin into

cracks less than 0.005 inch in width, while it is very difficult to retain injected epoxy resin in cracks greater than 0.25 inch in width, although high viscosity epoxies have been used with some success.  Epoxy resins cure to form relatively brittle materials with bond strengths exceeding the shear or tensile strength of the concrete.  If these materials are used to rebond cracked concrete that is subsequently exposed to loads exceeding the tensile or shear strength of the concrete, it should be expected that the cracks will recur adjacent to the epoxy bond line.  In other words, epoxy resin should not be used to rebond "working" cracks.

Epoxy resins will bond with varying degrees of success to wet concrete, and there are a number of special techniques that have been developed and used to rebond and seal water leaking cracks with epoxy resins.  These special techniques and procedures are highly technical and, in most cases, are proprietary in nature.  They may have application on Reclamation projects, but only after a thorough analysis has been performed to ensure that the more standard repair procedures will not be successful or cost effective.

*(b) Polyurethane Resins.*—Polyurethane resins are used to seal and eliminate or reduce water leakage from concrete cracks and joints. They can also be injected into cracks that experience some small degree of movement. Such systems, with the exception of the two-part solid polyurethanes, have relatively low strengths and should not be used to structurally rebond cracked concrete. Cracks to be injected with polyurethane resin should not be less than 0.005 inch in width. No upper limit on crack width has been established for the polyurethane resins at the time this is being written.

Polyurethane resins are available with substantial variation in their physical properties. Some of the polyurethanes cure into flexible foams. Other polyurethane systems cure to semiflexible, high density solids that can be used to rebond concrete cracks subject to movement. Most of the foaming polyurethane resins require some form of water to initiate the curing reaction and are, thus, a natural selection for use in repairing concrete exposed to water or in wet environments. At the time this is written, there are no standard specifications for polyurethane resins equivalent to the Standard Specification for Epoxy-Resin-Base Bonding Systems for Concrete, ASTM Designation C-881. This current lack of standards, combined with the wide variations possible in polyurethane physical properties, creates the necessity that great care be exercised in selecting these resins for concrete repair. "Cookbook" type application of these resins will not be successful. The Materials Engineering and Research Laboratory (D-8180) of the Denver Technical Services Center is currently testing and evaluating these very useful resin systems. They will provide advice and guidance for field applications if requested.

Because of the high costs (generally about $200.00 per linear foot of injected crack), resin

injection is not normally used to repair shallow, drying shrinkage, or pattern cracking.

The *Standard Specifications for Repair of Concrete, M-47* in appendix A contains current materials and procedures specifications for epoxy and polyurethane injection resins.

*(a) Preparation.*—Cracks, joints, or lift lines to be injected with resin should be cleaned to remove all the contained debris and organic matter possible. Several techniques have been used, with varying degrees of success, for cleaning such cracks. Once injection holes have been drilled, repeated cycles of alternately injecting compressed air followed by water have been very useful in flushing and cleaning cracks subject to water leakage. The successful use of soaps in the flushing water has been reported by some practitioners. Complete removal of such materials once injected into cracks is troublesome and may create more problems than it is worth. The use of acids to flush and clean cracks is not allowed by Reclamation. Cracks subject to epoxy injection for purposes of structural rebonding should not normally be injected with water. The epoxy resins will bond to wet concrete, but they develop higher bond strength when bonding to dry concrete.

*(b) Materials.*—Epoxy resin used for crack injection should be a 100-percent solids resin meeting the requirements of specification ASTM C-881 for type I or IV, grade 1, class B or C. If the purpose of injection is to restore the concrete to its original design load bearing capabilities, a type IV epoxy should be specified and used. If the purpose does not involve restoration of load bearing capabilities, a type I epoxy is sufficient. No solvents or unreactive diluents should be permitted in the resin.

Polyurethane resin used for crack injection should be a two-part system composed of 100-percent polyurethane resin as one part and water as the second part. The polyurethane resin, when mixed with water, should be capable of forming either a closed cell flexible foam or a cured gel, dependent on the water to resin mixing ratio. However, the resin should be such that, with appropriate water to resin mixing ratios, the resulting cured resin foam can attain at least 20-psi tensile strength with a bond to concrete of at least 20 psi and a minimum elongation at tensile failure of 400 percent. The manufacturer's certification that his product meets these minimum requirements should be required before the injection resins are accepted for use on the job.

*(c) Injection Equipment.*—Resins can be injected with several types of equipment. Small repair jobs employing epoxy resin can use any system that will successfully deposit the epoxy in the required zones. Such systems could use a prebatch arrangement in which the two components of the epoxy are batched together prior to initiating the injection phase with equipment such as small paint pressure pots. The relatively short pot life of the epoxy makes this technique rather critical as far as timing is concerned.

Large epoxy injection jobs generally require a single-stage injection technique in which the two epoxy components are pumped independently of one another from the reservoir to the mixing nozzle. At the mixing nozzle, located adjacent to the crack being repaired, the two epoxy components are brought together for mixing and injecting. The epoxy used in this injection technique must have a low initial viscosity and a closely controlled set time. Several private companies have proprietary epoxy injection systems (figure 62). These organizations have developed epoxies and techniques which allow them to make satisfactory repairs under the most adverse conditions. One or more of these companies should be contacted regarding any major repairs requiring the epoxy injection technique. Names and addresses of these companies can be obtained from the Materials Engineering and Research Laboratory, Code D-8180, Denver, Colorado.

Polyurethane resins have a very short pot life after mixing and are always prepared and

injected with *multiple component*, single stage proprietary equipment similar to that used for large scale epoxy repairs. Reclamation specifications do not permit single component injection of 100-percent pure resin. In every instance, multiple component water-resin mixtures or resin (part A) -resin (part B) mixtures must be used. This equipment mixes the resin system components just prior to the point of crack injection. The size of polyurethane injection equipment varies from small, hand operated, pumps to full size commercial equipment capable of discharging many cubic feet of resin per hour (figures 63 and 64). The pumping pressure required of polyurethane injection equipment may exceed 3,000 psi. There are a number of manufacturers of high quality polyurethane resin injection equipment, and there is seldom any cause to attempt polyurethane injection on a Reclamation project with equipment designed for, or adapted from, other operations. Such adaptation is usually indicative of an inexperienced contractor and is highly discouraged.

*(d) Application.*—The success of resin injection repair projects is directly related to the experience and knowledge of the injection contractor. Reclamation requires that an injection contractor have a minimum of 3 years' experience in performing injection work similar to that being contracted for and that a minimum of five projects be included in that experience. Reclamation may also accept an injection contractor not having the required experience provided that the work is performed under the full-time, direct technical supervision of the injection resin manufacturer, provided the manufacturer has a minimum of 5 years' experience providing resins for applications similar to those specified.

*(1) Application of Epoxy Resin by Pressure Injection.*—The objective of epoxy resin injection is to completely fill the crack or delamination being injected and retain the resin in the filled voids until cure is complete. The first step in the resin injection process is to thoroughly clean the concrete surface in the

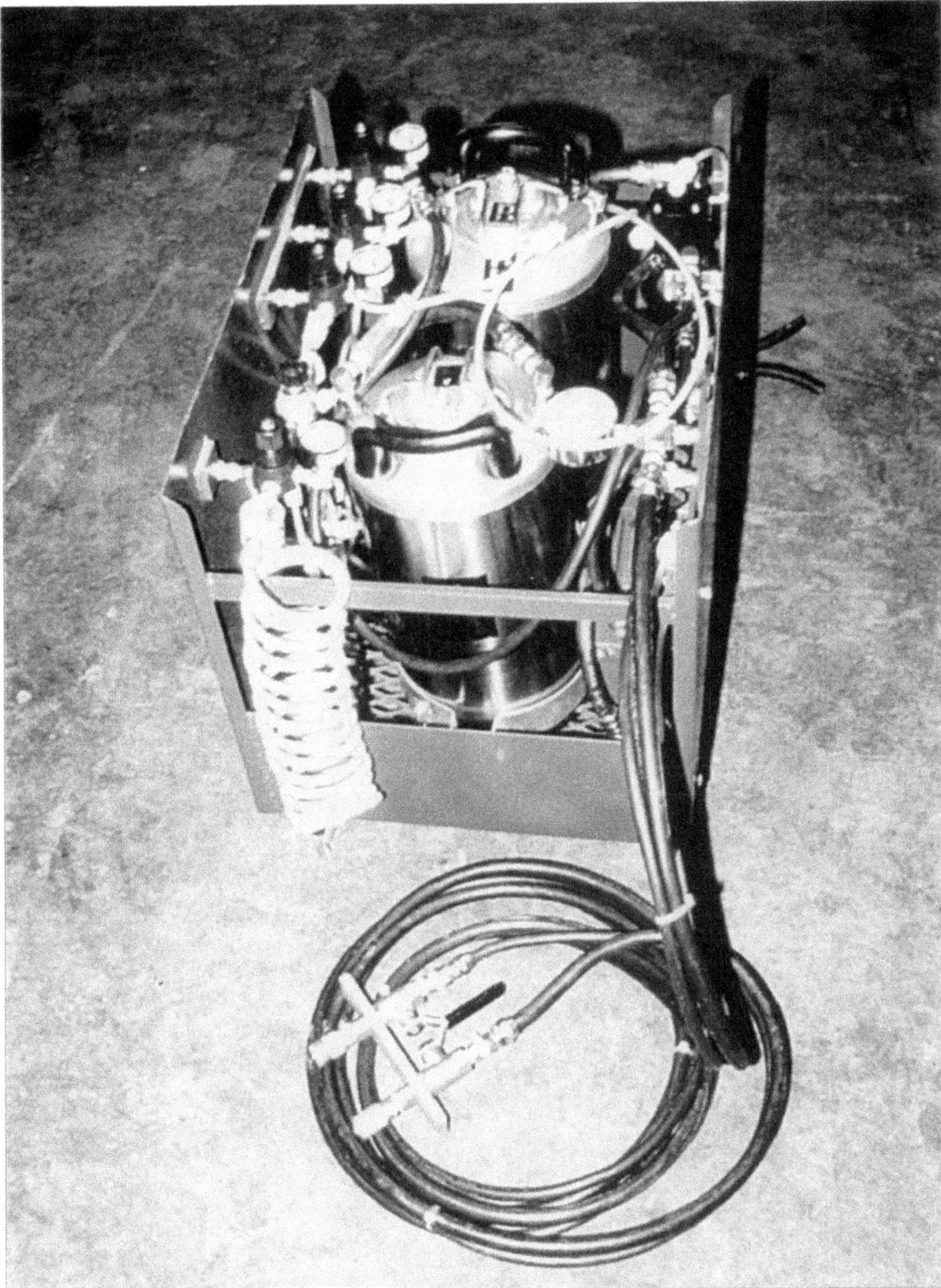

Figure 62.—Proprietary epoxy injection equipment. Such equipment does not mix resin components until the point of injection.

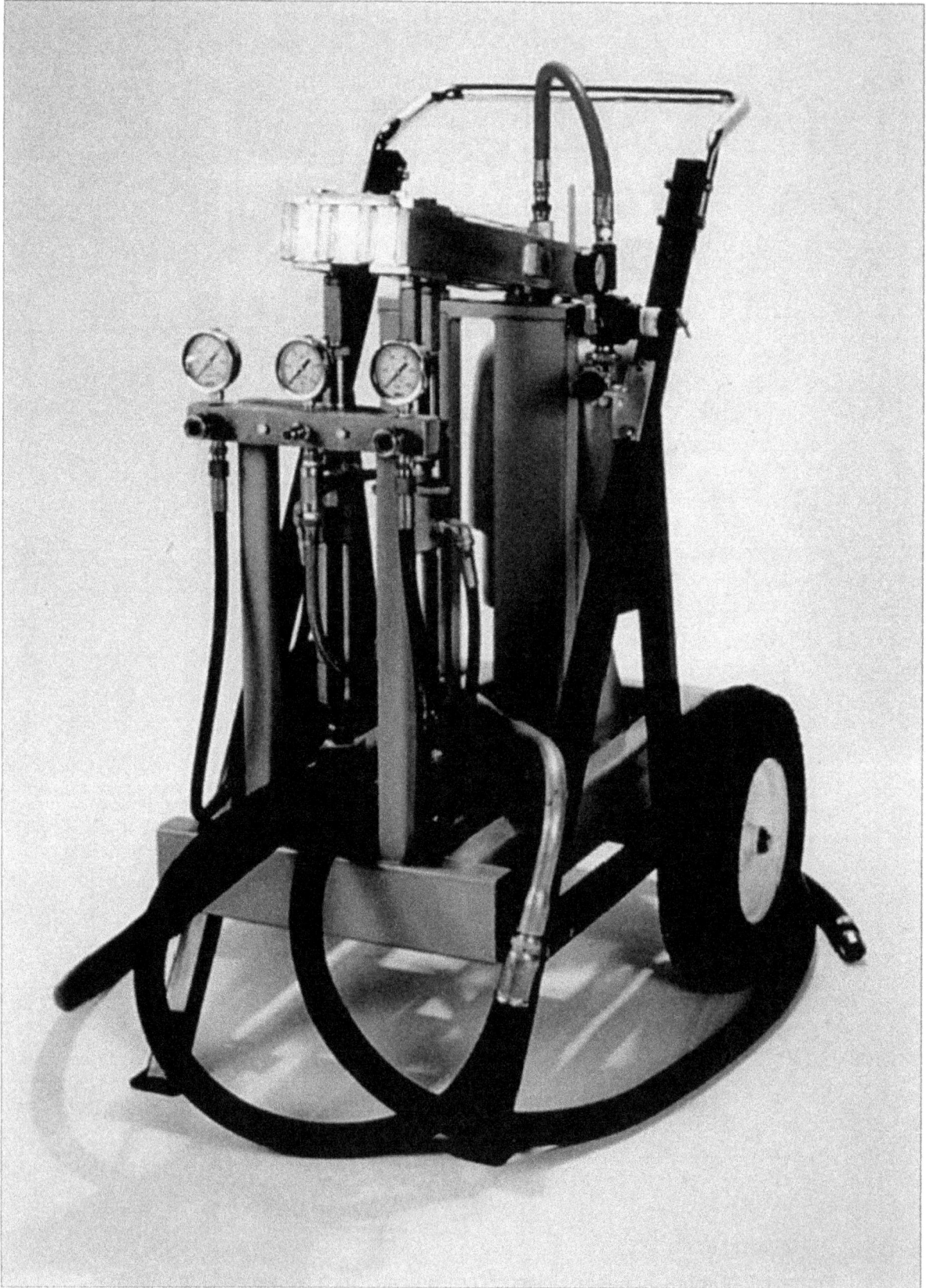

Figure 63.—Commercial polyurethane injection pump.

Figure 64.—This is an air-powered pump system used for large scale polyurethane resin injection.

vicinity of the cracks of all loose or deteriorated concrete and debris. The area of injection is then inspected and the injection port location pattern established. Several different types of injection patterns can be used:

- If the cracks are clearly visible and relatively open, injection ports can be installed at appropriate intervals by drilling directly into the crack surface. Care should be taken in drilling the ports to prevent drilling debris and dust from blocking or sealing the openings. Special vacuum drill chucks are available for this work. The surface of the crack between ports is then sealed with epoxy paste and the paste is allowed to cure. Epoxy injection begins at the lowest elevation port and proceeds along and up the crack to the uppermost port.

- A more positive method is to drill holes on alternate sides of the crack, angled to intersect the crack plane at some depth below the surface. This method ensures that the crack will be intersected even if

it strikes or dips in unexpected directions. The top surface of the crack is then sealed with epoxy paste, and injection is accomplished as described above.

Low to moderate epoxy injection pressures should be used and patience should be exercised to permit the resin to flow and completely fill the voids existing in the concrete. The use of high injection pressures can result in flow blockage and incomplete filling and, generally, is an indication of an inexperienced contractor.

The best method of ensuring quality epoxy injection work is to require the contractor to prepare and submit for approval his overall, detailed injection plan and then to obtain small diameter proof cores from the injected concrete. If more than 90 percent of the voids in the cores are filled with hardened epoxy, the injection can be considered complete. If injection is not complete, the contractor should be required to reinject the concrete and obtain additional cores at no additional cost to the Government.

*(2) Application of Polyurethane Resin by Pressure Injection.*—The basic procedure for polyurethane injection consists of first gaining control of the leaking water, followed by pressure injecting resin to seal the cracks. In most instances, the polyurethane injection procedure is almost identical to the processes followed for cementitious grouting.

To gain control of the waterflow, holes are drilled to intercept the waterflow paths as far as possible from the concrete surface. Valved drains known as "wall spears" (figure 65) are installed in the drilled holes, opened, and used to relieve water pressure in the cracks near the surface. The cracks are then temporarily sealed with wood wedges, lead wool, or resin soaked jute rope to prevent excessive loss of injection resin.

Additional resin injection holes are then drilled on alternate sides of the crack at a maximum spacing of 24 inches. These holes are angled to intercept the crack at a depth of 8 to 24 inches (as concrete thickness allows, these holes should extend as deeply as possible). Injection ports of various design (figure 66) or additional valved wall spears may be installed in the drilled holes, depending on the injection plan and the presence of flowing water.

Polyurethane resin injection should occur according to a preplanned sequence. A system of split spacing similar to cementitious grouting is often successful. In such a system, the primary holes are injected first, followed by drilling and injection of secondary holes located between the primary holes. Similarly, tertiary holes, located between the secondary holes and primary holes are then drilled and injected. Injection pressures should be the minimum pressures necessary to accomplish resin travel and filling. Even so, pressures of 1,500 to 2,000 psi are common in this work. Closure of each injection hole should be accomplished by holding injection pressure for a period of 10 to 15 minutes after injection flow has ceased. This technique of "closure to absolute refusal" ensures that the resin attains

maximum density in the crack and becomes a permanent repair. It is usually a mistake to stop injection as soon as the water leakage is stopped. If such a procedure is followed, the partially cured, low density resin can be pushed out of the crack system by hydrostatic pressure, and repeat injection will be required to seal the resulting leakage.

It is also common practice to intermittently inject resin into a port in order to accomplish sealing of large waterflows. With this technique, a preselected quantity of resin is slowly injected into a port, followed by a 15-minute to 2-hour waiting period before repeat injection. Several such cycles of injection may be necessary to control and seal large waterflows. It is still necessary that closure to absolute refusal be accomplished with the final injection cycle.

Polyurethane resin injection is accomplished with varying water to resin ratios. In cases of high waterflows, it may be desirable to inject water to resin ratios as low as 0.5:1. Alternatively, the water and resin may be introduced and mixed in a "residence tube" 1 to 5 feet before the point of injection so the foaming reaction may be well underway upon entering the crack network. Special downhole packers can be utilized to inject resin at points deep within a structure. If resin components are mixed and injected at the surface of such deep holes, the reaction will occur within the drill hole before reaching the desired point of injection into the cracks. These special packers (figure 67) allow separation of the resin components until they reach the downhole point of crack injection.

The necessity of using experienced injection contractors or technical advisors for work of this nature cannot be overemphasized.

*(3) Cleanup.*—At the completion of resin injection, all injection ports, excess resin, and crack surface sealer should be removed from surfaces that are visible to the public. This can be accomplished by scraping, high pressure

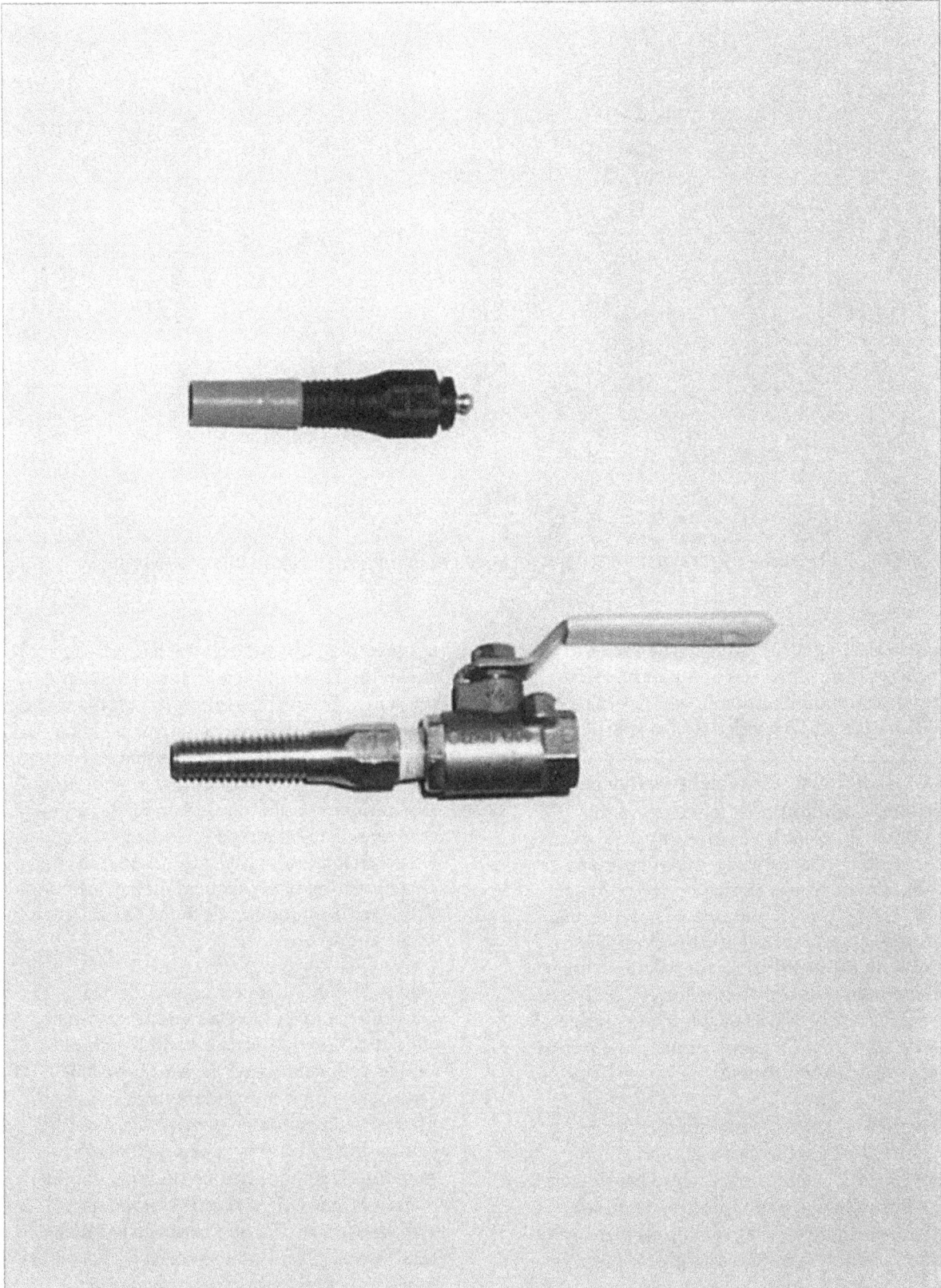

Figure 65.—An injection port with zirc fitting and a valved wall spear are shown in this photograph. The wall spear can be used to relieve water pressure and to inject resin.

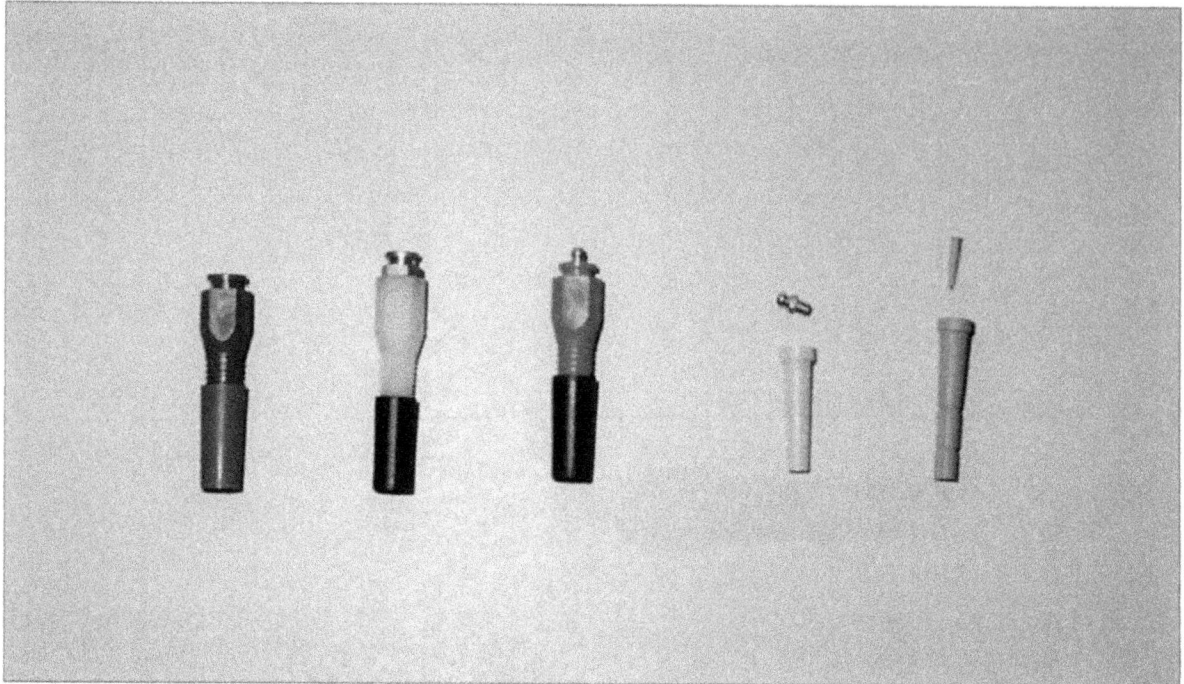

Figure 66.—Several different types of injection ports are shown in this photograph.

water blasting, or grinding. The use of dry pack or other replacement repair material necessary to fill injection holes should be anticipated and provided by the specifications.

**35. High Molecular Weight Methacrylic Sealing Compound.**—Concrete sealing compounds (see also section 38) are applied to cured, dry concrete as a maintenance procedure to reduce or prevent penetration of water, aggressive solutions, or gaseous media and the associated deterioration, such as freeze-thaw, carbonation, or sulfate damage. These materials replace the linseed oil based treatment, which was generally misunderstood and is no longer recommended for use on Reclamation concrete.

A variety of different membrane forming (similar to paints or coatings) and surface penetrating chemicals are manufactured and sold as sealing compounds for concrete surfaces. Some of these materials provide very good protection to the concrete for discrete periods of time. Other commercially available sealing materials, however, may be little more than mineral spirits and linseed oil. Such

systems will, at best, do little harm to the concrete. Their application may, however, prevent subsequent treatment with the sealing compounds that have been proven effective. For this reason, only products that have proven effective in standardized laboratory evaluations should be used on Reclamation concrete. The Materials Engineering and Research Laboratory, D-8180, maintains a current listing of concrete sealing compounds that have been found effective for Reclamation applications.

One type of sealing compound that has proven effective in Reclamation laboratory tests and field applications and has been designated a Standard Repair Material is known as a high molecular weight methacrylic monomer system. This sealing compound is composed of a methacrylic monomer and appropriate polymerization "catalysts" very similar to the monomer system used in polymer concrete (section 32). It is a water thin, amber colored liquid that is easily spread over horizontal and vertical concrete surfaces with brooms or squeegees.

Figure 67.—A proprietary downhole packer allows separation of the resin components until they reach the downhole point of injection.

The liquid penetrates the concrete surface to a depth of about 1/16 inch *but is most effective in penetrating and sealing cracks in the concrete surface.* This sealer will act like a membrane forming system if excess monomer is applied or if two or more applications are made. The appearance of the concrete following application will be somewhat like a varnished or water wet surface and may be splotchy in areas of high and low absorption. Cured sealer left on the surface of the concrete will be deteriorated by solar radiation within 1 to 2 years and will disappear. The loss of this surface material is of no consequence since the objective of the application is to penetrate and seal cracks where the sealer is protected from solar radiation deterioration. The expected service life of properly applied methacrylic sealing compound under typical Western State climatic conditions is 10 to 15 years. Reapplication is then necessary. Figure 68 shows application of a high molecular weight methacrylic sealing compound to the crest of Kortes Dam.

*(a) Preparation.*—Concrete to receive methacrylic sealing compound must be dry, clean, and physically sound. The cracks and porosity of wet or damp concrete will be completely or partially filled with water that will prevent the desired penetration of the sealing compound. The concrete is suitably dry if no moisture appears under a sheet of transparent polyethylene taped to the concrete surface during a minimum 2-hour exposure to full sunlight.

Power sweeping or hand brooming followed by blowing with high pressure compressed air should be used to remove all debris from the surface. Very small areas of paint, asphalt, rubber, or similar type coatings can usually be ignored. It should be expected, however, that the methacrylic monomer system will attack and deteriorate most markings or coatings intentionally placed on the concrete. Deteriorated or unsound concrete should be removed following the methods described in

Figure 68.—High molecular weight methacrylic sealing compound is being applied to the crest of Kortes Dam, near Casper, Wyoming.

section 8. After proper preparation, the concrete must be protected from construction traffic and wetting.

*(b) Materials.*—The high molecular weight methacrylic sealing compound is usually obtained from the manufacturer as a three-component system:

(1) A water thin liquid methacrylic monomer

(2) Cuemene hydroperoxide initiator (or catalyst)

(3) Cobalt napthenate promoter

Each manufacturer will specify the proper proportions for mixing these materials, and these recommendations must be closely followed. As a general rule, it can be expected that the initiator will be added to the monomer at about 4 to 6 percent, by weight and the promoter at about 1 to 3 percent, by weight. *The initiator and promoter must never be directly mixed with each other, as this will produce an extremely violent and explosive chemical reaction.* Rather, the initiator is first thoroughly mixed with the monomer and then the promoter is added and mixed with the monomer-initiator mixture.

Some manufacturers supply a two-part monomer system already containing the proper proportion of promoter. With such systems, it is necessary to add only the proper quantity of initiator. Once all the components are mixed, the sealing compound system will have a definite and short pot life that cannot be extended under normal conditions.

For this reason, if the project is of sufficient size to require more than 5 gallons of sealing compound, it is usually best to premeasure the required quantity of monomer into separate 5-gallon volumes which are not mixed but are placed at appropriate locations along the application path; each location consists of 5 gallons of monomer, the appropriate quantity of initiator, and the separate and appropriate quantity of promoter (if not already added to the monomer by the manufacturer). During application, as one 5-gallon volume of sealing compound is nearly used up, an additional 5 gallons can then be mixed for use. Since all the 5-gallon materials quantities are premeasured and located along the application path, this system also helps ensure that all necessary components are added and mixed for each volume of sealing compound. Otherwise, it is somewhat common to discover that either the initiator or the promoter has been omitted from sealing compound already applied to the concrete, thereby creating a bothersome situation.

*(c) Application.*—As the mixed monomer system has a distinctly short pot life at normal ambient temperatures, it is important not to mix more material than can be easily applied prior to the system becoming too thick—normally within 15 to 20 minutes. If the material is being applied by two workmen and the application path is clear and easily attained, it would be common to mix up to 5-gallon quantities of material unless the ambient temperatures exceed 85 °F. The sealant should be applied immediately after mixing. Application of the sealant system to concrete in direct sunlight should be avoided. Solar radiation greatly shortens the working pot life of the mixed sealant. Workmen can use squeegees, industrial size paint rollers, push brooms, brushes, or airless powered spray systems to apply the sealant system. What is desired is to flood the concrete with a heavy uniform coverage without leaving puddles of excess material. Application rates will normally vary from 75 to 100 square feet per gallon, depending on surface roughness and absorption. Vertical surfaces should receive two or more brushed, sprayed, or rolled applications. Repeat applications should be made immediately, without waiting for the sealant to cure.

If a skid-resistant surface is required for foot or vehicle traffic, sand must be broadcast over the liquid sealant system within 15 to 20 minutes of application. The sand gradation is contained in the appendix A specifications. The sand application rate should be about 1/4 to 1/2 pound per square yard of surface.

*(d) Curing and Protection.*—After application, the treated surfaces should be protected for a minimum of 24 hours. Protection should be provided for 48 to 72 hours if ambient temperatures are lower than 50 °F to permit the sealant to fully cure. Sealant applied during the night will be quickly cured by solar radiation the following morning. Night applications allow the maximum penetration into cracked surfaces and should be pursued whenever practicable.

*(e) Safety.*—The safety provisions of section 33.(e) should be followed when using high molecular weight methacrylic sealants. Storage of initiator and promoter in the same room is prohibited. Storage of initiator under refrigerated conditions is desirable. Every precaution should be taken to prevent mixing or direct contact of initiator and promoter.

**36. Polymer Surface Impregnation.**—The polymer surface impregnation process was developed by Reclamation for the Federal Highway Administration to prevent chloride deicing salt penetration and subsequent corrosion of reinforcing steel in existing concrete highway bridge decks. The process has provided in excess of 20 years of highly successful protection to many highway bridge decks and was applied to the entire roadway surface over Reclamation's Grand Coulee Dam.

In this process, the concrete to be treated is first covered with a bed of sand and dried with heat to remove moisture from the zone to be impregnated. A low viscosity methyl methacrylic monomer system is applied to the sandbed under a heavy polyethylene sheet and allowed to soak into the concrete surface for about 6 hours. The polyethylene retards evaporation of the highly volatile monomer system, and the sand acts as a reservoir, retaining the monomer system on the concrete until it soaks into the surface. The polyethylene covered sand and treated surface is then reheated to initiate in-situ polymerization of the methyl methacrylate monomer system within the structure of the concrete. Concrete

so treated will be virtually impervious to water absorption and freeze-thaw deterioration.

Detailed materials and performance specifications for the polymer surface impregnation process are contained in section 3.13 of appendix A. Users considering application of this procedure should carefully appraise the current costs and associated safety issues. The costs of energy to properly dry and reheat areas of concrete may preclude large scale use of this very effective preventative maintenance process.

**37. Silica Fume Concrete.**—Silica fume concrete is conventional portland cement concrete containing admixtures of silica fume. Silica fume is a finely divided powder by-product resulting from the use of electric arc furnaces. When mixed with portland cement concrete, silica fume acts as a "super pozzolan." Concrete containing 5 to 15 percent silica fume by mass of cement commonly can develop 10,000- to 15,000-psi compressive strengths, reduced tendency to segregate, very low permeabilities, and enhanced freeze-thaw and abrasion-erosion resistance. Reclamation use of silica fume is primarily for the purpose of enhancing or improving concrete durability with less emphasis on strength improvement.

Silica fume concrete is the repair material of choice for applications requiring enhanced abrasion-erosion resistance and/or reduced permeability. Silica fume concrete requires a very thorough curing procedure, however, and should not be used unless such a procedure can be accomplished. Otherwise, this repair material is used in accordance with the provisions for conventional replacement concrete.

The silica fume admixture can be obtained in at least three forms for use in concrete:

(1) Silica fume powder

(2)  Densified silica fume powder with or without a high range water reducing admixture (HRWRA) and other dry admixtures

(3)  Silica fume-water slurry with HRWRA and other admixtures

The dry silica fume powder is difficult to use because of its extremely small particle size. The fine powder drifts and spreads in any draft or air movement and can create silicate respiratory problems to workers. The water slurry form is easy to use, creates no dust problems, but does involve an additional mixing step as the slurry settles during storage and shipment and must be stirred and remixed prior to use. The water in the slurry must be accounted during mix design. Transportation costs of the slurry must also be considered. The densified powder form is currently the most convenient form to ship and use. In this form, the silica fume powder is compacted and densified and does not produce nearly the quantity of dust that occurs with the powdered form. It is a dry material and does not create additional shipping costs. Because it has been compacted into clumps, it should be expected that additional time would be required during mixing to fully break up and disperse the densified silica fume admixture in the concrete mixture.

Because of these various forms, it is essential that trial mixtures be prepared and tested during mix design to ensure development of the desired concrete properties.

The addition of silica fume admixture to concrete will increase the water requirement due to the high surface area of the very fine silica fume particles. The use of HRWRA is thus necessary to obtain the maximum strength and durability with silica fume concrete. Provisions should be made during proportioning, however, to accommodate the slump gain commonly associated with concrete containing HRWRA. Silica fume increases the cohesion or "stickiness" of the concrete and can result in workability and finishing problems for those inexperienced in the proper finishing techniques. It is of primary importance to place and finish silica fume quickly, before excessive mix water evaporation and stiffening occurs. The slump gain from the HRWRA commonly offsets some of the "stickiness" of silica fume concretes. Reclamation requirements include 4-6 percent entrained air in silica fume concrete. This addition of air entraining admixture also improves the workability of the concrete.

Special repair techniques are required for restoration of damaged or eroded surfaces of spillway or outlet works tunnel inverts and stilling basins. In addition to the usual forces of deterioration, such repairs often must withstand enormous dynamic and abrasive forces from fast-flowing water and sometimes from suspended solids. The enhanced abrasion-erosion resistance and high strength of silica fume concrete makes it the repair material of choice for these types of repair. It should be recognized, however, that the cause(s) of the original damage in such repairs must be mitigated if a permanent repair is to be accomplished.

Whenever practicable, low slump silica fume concrete should be used for these types of repair. Slump of the concrete should not exceed 2 inches for slabs that are horizontal or nearly horizontal and 3 inches for all other concrete. (This is 1 inch less slump than required in the M-47 specifications for conventional applications of silica fume concrete.) The net water-cementitious ratio (exclusive of water absorbed by the aggregates) should not exceed 0.35, by weight. An air-entraining agent should be used, and a high range water reducing admixture should be used. Set-retarding admixtures should be used only when the interval between mixing and placing is quite long.

*(a) Preparation.*—Concrete to be repaired with silica fume concrete should be prepared in accordance with the requirements of section 8.

*(b) Materials.*—There are currently no ASTM specifications for the silica fume admixture.

Reclamation has been approving silica fume meeting the requirements of the Standard Specification for Microsilica for Use in Concrete and Mortar, AASHTO Designation: M 307-90.

All other materials, including the requirement for 4- to 6-percent entrained air, should be in accordance with the provisions of section 29.

*(c) Application.*—Silica fume concrete must be mixed, transported, and placed in accordance with the highest quality procedures for conventional concrete technology and with the provisions of section 29.

The following silica fume concrete mix design is included as a reference or starting point only. Proper proportioning of silica fume concrete requires trial mixes:

Material

| | |
|---|---|
| Portland cement | 679.3 lb/yd³ |
| Coarse aggregate (3/4-in. MSA) | 1,910.9 lb/yd³ |
| Sand | 1,036.2 lb/yd³ |
| Water | 169.0 lb/yd³ |
| Silica fume slurry¹ = 13.3 percent by mass of cement | 200.6 lb/yd³ |
| Total | 3,996.0 lb/yd³ |
| HRWRA | 300 cc |

w/c ratio = $\frac{\text{Water} + 0.51(\text{slurry})}{\text{Cement} + 0.45(\text{slurry})} = \frac{239.2}{769.5} = 0.35$

¹ Silica fume slurry contains 51 percent water, 45 percent cementitious silica fume, and 4 percent admixtures, all by mass. The admixtures consist of HRWRA and air entraining admixture. In this instance, additional HRWRA was required to produce a workable concrete.

Silica fume concrete is placed and finished using conventional equipment and procedures. As previously discussed, placement should be done quickly to allow finishing before stiffening occurs. Consolidation and compaction should be accomplished with internal vibrators. Vibrating screeds can be used for larger placements and, usually, a single pass of multirow screeds will provide an adequate surface finish. Small repair areas can be hand screeded (figure 69). Bull floats can be used after screeding, but floating must be done immediately after screed passage (figure 70). Hand troweling silica fume concrete is usually not too effective, except for small repairs, because it takes too long. There is very little bleed water development after placing silica fume concrete. This can result in plastic and drying shrinkage cracking under conditions of elevated temperatures, low humidity, and high wind conditions which cause rapid water evaporation from the concrete surface (see section (d) below). The use of a long chain cetyl alcohol evaporation retarding finishing aid is highly recommended under such conditions.

*(d) Curing.*—Silica fume concrete must be properly cured if a successful repair is to result. Fresh silica fume concrete has very low or no bleed water development. This is due to the affinity silica fume has for water and to the low water-cement ratio of the mix. If the rate of evaporative water loss from the surface of freshly placed silica fume concrete exceeds the rate of bleed water development, plastic shrinkage cracking of the surface will result. Evaporative water loss must be prevented by such measures as immediate application of curing compound, covering the fresh concrete with plastic membrane, water fogging or flooding, use of an evaporation retarding finishing admixture, and by immediate curing. The common practice (with conventional concrete) of allowing the development and evaporation of bleed water from the surface prior to beginning curing will always result in cracking of silica fume concrete. It is best to begin the curing of silica fume concrete immediately after finishing. Very successful curing of stilling basin repairs has resulted from flooding the repair area with water immediately after the silica fume concrete has attained initial set. Even so, water evaporation from the concrete surface must be prevented prior to flooding. The spillway repairs shown in figures 69 and 70 were cured by applying curing compound and covering with polyethylene immediately after floating, even while placing operations were continuing (figure 71). Water trickler hoses were placed

Figure 69.—This workman is hand screeding a small silica fume concrete repair.

Figure 70.—Using a bull float on a silica fume repair. Finishing must be done immediately after screeding.

Figure 71.—Curing compound and polyethylene sheeting should be applied to cure silica fume concrete as soon as finishing is completed if drying shrinkage cracking is to be prevented.

under the polyethylene as soon as the concrete attained initial set, and water was applied continuously for 14 days. This thin, 4- to 6-inch deep, repair has experienced virtually no shrinkage cracking.

*(e) Safety.*—The same safety provisions should be followed as with the use of any portland cement product, except that additional precautions must be taken to prevent inhalation of silica fume dust. The high silica content of silica fume dust can lead to development of silicosis if proper respirators are not used by workmen participating in or downwind of mixing operations.

**38. Alkyl-Alkoxy Siloxane Sealing Compound.**—This sealing compound is effective in reducing water penetration into treated concrete, provided that the sealing compound contains in excess of 15-percent siloxane solids. These solids are suspended in a carrier such as alcohol or mineral spirits that evaporates from the concrete following application. Use of this type of sealing

compound does not cause a change in the appearance of the treated concrete, except that the darkening normally associated with water wetting of concrete does not occur. It is common to see water bead on treated concrete surfaces. Siloxane sealing compounds will not provide protection to concrete that is completely inundated in water except for short periods. They should not be used under conditions that involve prolonged inundation such as occurs with stilling basins, canal floors, or spillway floors unless there are significantly long dry periods between inundations and it is acceptable to reapply the sealing compound following inundation. Siloxane sealing compound is best used to protect concrete from rain, snowmelt, and water splash zones.

These materials have a relatively limited service life, and reapplication should be scheduled about every 5 to 7 years for optimum protection. Application of these materials, however, proceeds very quickly on horizontal and vertical concrete surfaces. Two workmen can be expected to treat 10,000 to

15,000 square feet of horizontal surface in a day. Cost of siloxane sealing compound and application average about $0.50 to $0.70 per square foot.

*(a) Preparation.*—Concrete surfaces to be treated with siloxane sealing compound should be dry, clean, and sound. All deteriorated concrete should be removed (section 8), and all needed repairs should be accomplished prior to application. As a general rule, power sweeping and, in some instances, high pressure (less than 8,000 psi) water washing of the surfaces will be sufficient. The concrete should be allowed to dry for 24 to 48 hours following wetting. Cracks in the concrete should be cleaned and blown out with compressed air to remove any debris. Once preparation (including drying if needed) is completed, the sealing compound should be applied within 24 hours.

*(b) Materials.*—Siloxane sealing compound should be a one-part, ready to apply, oligomeric alkyl-alkoxy siloxane containing no less than 20 percent active solids, by mass, in a clear organic carrier. Siloxane sealing compounds containing less than 20 percent active solids do not provide sufficient protection to the concrete and are not allowed by section 3.15.d of the M-47 specifications in appendix A. The material should be supplied in 5-gallon pails or 55-gallon drums. No additional mixing or blending should be necessary prior to application. No additional solvents or diluents should be added to the supplied sealing compound.

*(c) Application.*—The sealing compound should be applied to the concrete using squeegees, push brooms, paint rollers (figure 72) or low pressure airless spray equipment. Adjacent glass, metal, and painted surfaces must be protected from the sealing compound. Overspray protection must be provided if spray application equipment is utilized. Fine spray mists can travel extensively and damage downwind structures, equipment, and automobiles.

The application rate should be 1 gallon per 80 to 120 square feet of concrete surface. Horizontal and vertical surfaces should receive two wet to wet coatings, 5 or so

Figure 72.—A paint roller application of siloxane sealing compound to the downstream face of Nambe Falls Dam, near Santa Fe, New Mexico.

minutes apart, flooding the surface each time. Excess sealant should be broomed out until it is absorbed by the concrete. Low density areas of the concrete surface need the most protection, and these areas are the most absorptive.

*(d) Curing.*—Siloxane treated concrete must be protected from foot and vehicular traffic and water wetting for 24 hours following application. Should the concrete experience rain or heavy water splashing during this period, the drying step of preparation should be repeated and reapplication made.

*(e) Safety.*—Siloxane sealing compounds are formulated with organic solvents such as alcohol or mineral spirits. Materials Safety Data Sheets should be consulted concerning flammability and toxicity. Applicators should wear protective coveralls, rubber boots, eye protection, and approved respirators.

# Chapter V

# Nonstandard Methods of Repair

## 39. Use of Nonstandard Repair Methods.—

The standard concrete repair methods/ materials discussed in chapter IV will meet nearly all concrete repair needs. There will be occasions, however, resulting from unusual causes of damage and exposure conditions or special repair needs, when the standard repair methods may not meet the performance needs. In these infrequent instances, non-standard repair methods may be required.

Repair materials are considered to be non-standard if they have not been thoroughly tested and evaluated for Reclamation applications. The use of such materials involves a certain element of risk because the performance properties of these materials are unknown or not fully defined. The application of nonstandard materials can be justified only when it has been determined that no standard repair material will serve, and if all parties associated with or responsible for accomplishing the repairs are made to understand the risks and agree to accept the uncertainties and possible consequences.

An example of such a situation would be the need to repair concrete damage on the crest of a dam that is less than 1-1/2 inches deep. There is no standard repair material suitable for such shallow repairs when exposed to sunlight temperature variation conditions. (These conditions eliminate the use of epoxy mortars.) The current standard repair material for depths between 1-1/2 inches and 6 inches is epoxy-bonded replacement concrete. If the deterioration is not at least 1-1/2 inches deep, sufficient concrete must be removed to accomplish a 1-1/2- inch depth. If, for some reason, it was undesirable or impossible to remove the required depth of concrete, it might be appropriate to select one of the proprietary thin repair products. These

materials have been only partially evaluated (Smoak and Husbands, 1996) and the long-term field performance has not yet been fully determined. In this example, the people responsible for the dam must be contacted, and the risks of using an unproven repair material must be made known to them. These responsible personnel, with this knowledge, can then determine if the benefits of performing the shallower repairs would outweigh the uncertainties associated with unproven performance.

*(a) Preparation.*—Concrete preparation for repair with nonstandard materials should be in accordance with section 8 unless required otherwise by the materials manufacturer. The recommendations of the manufacturer should be closely followed.

*(b) Materials.*—Nonstandard repair materials will be proprietary or commercial products in most instances. Manufacturer's technical representatives should be contacted and made fully aware of the nature of the problem and the causes for selecting nonstandard repair materials. They must be informed of any secondary considerations, such as the need to accomplish repairs on an emergency basis or the inability of taking the structure to be repaired out of service except at certain times. Copies of materials warranties should be obtained if available. Technical representa-tives should be questioned closely about the conditions under which their materials *would not be suitable* for use or when the materials *warranties would not apply*. The manufacturer should be able to supply case studies or project histories of use of their materials to repair similar damage.

*(c) Application.*—Application of nonstandard repair materials must be made in exact conformance to the manufacturer's

recommendations. Failure to apply the materials properly will void any warranties and is a commonly cited reason for proprietary material failure. Construction inspection reports should closely note any contractor variance from the manufacturer's recommended application procedures.

*(d) Curing.*—Curing of nonstandard repair materials must be performed exactly as recommended by the materials manufacturer.

*(e) Safety.*—Nonstandard repair materials could contain compounds not commonly encountered during repair construction. Copies of the Materials Safety Data Sheets must accompany the shipment of materials to the jobsite. They should be consulted and appropriate safety provisions developed.

American Concrete Institute, "Guide for Evaluation of Concrete Structures Prior to Rehabilitation," ACI 364.1R-93, Detroit, Michigan, September 1993.

American Concrete Institute, "Guide for the Use of Preplaced Aggregate Concrete for Structural and Mass Concrete Applications," ACI 304.1R-92, Detroit, Michigan, 1992.

American Concrete Institute, "Building Code Requirements for Reinforced Concrete," ACI 318R-92, Detroit, Michigan, 1992.

American Concrete Institute, "Cement and Concrete Technology," ACI 116R-90, Detroit, Michigan, 1990.

American Concrete Institute, "State of the Art Report on Fiber-Reinforced Shotcrete," ACI 506.1R-84, Detroit, Michigan, 1984.

American Concrete Institute, "Specification for Materials, Proportioning, and Application of Shotcrete," ACI 506.2-77, revised 1983, Detroit, Michigan, 1977.

American Concrete Institute, "Recommended Practice for Shotcreting," ACI 506-66, revised 1983, Detroit, Michigan, 1966.

Bureau of Reclamation, "Standard Specifications for Repair of Concrete," M-47 (M0470000.296), Technical Service Center, Denver, Colorado, February 1996.

Bureau of Reclamation, "Concrete Manual," Eighth Edition, Technical Service Center, Denver, Colorado, 1975.

Bureau of Reclamation, "Alkalies in Cement and their Effect on Aggregate and Concretes," Report No. Ce40, Technical Service Center, Denver, Colorado, July 1, 1942.

Emmons, Peter, "Concrete Repair and Maintenance Illustrated," R.S. Means Company, Inc., PO Box 800, Kingston, Massachusetts, 1994.

Poston, R.W., A. Rhett Whitlock, and Keith E. Kesner, "Condition Assessment Using Non-destructive Evaluation," Concrete International Magazine, Volume 17, No. 7, American Concrete Institute, Detroit, July 1995.

Price, Walter H., "In the Beginning," Concrete Laboratory Technical Conference, Concrete and Structural Branch, Division of Research, Bureau of Reclamation, December 2-3, 1981

Smoak, W. Glenn and Tony B. Husbands, "Results of Laboratory Tests on Materials for Thin Repair of Concrete Surfaces," Technical Report No. REMR-CS-52, U.S. Army Corps of Engineers, Waterways Experiment Station, Vicksburg, Mississippi, September 1996.

Smoak, W. Glenn, "Effect of High Range Water Reducers on Cement Grout," Concrete International Magazine, Volume 15, No. 1, American Concrete Institute, Detroit, Michigan, January 1993.

Smoak, W. Glenn, "Repairing Abrasion-Erosion Damage to Hydraulic Concrete Structures," Concrete International Magazine, Volume 13, No. 6, American Concrete Institute, Detroit, Michigan, June 1991.

Stark, David and G.W. DePuy, "Investigations of Alkali-Silica Reactivity in Five Dams in the Northwestern United States," Report No. R-95-05, Bureau of Reclamation, Technical Service Center, Denver, Colorado, March 1995.

Travers, Fred, "Acoustic Monitoring of Prestressed Concrete Pipe at the Agua Fria River Siphon," Bureau of Reclamation, Technical Service Center, Denver, Colorado, December 1994.

U.S. Army Corps of Engineers, "Evaluation and Repair of Concrete Structures," Engineer Manual, EM 1110-2-2002, Washington, DC, June 1995.

Warner, James, "Understanding Shotcrete - The Fundamentals," Concrete International Magazine, Volume 17, No. 5, American Concrete Institute, Detroit, Michigan, May 1995.

# Standard Specifications for the Repair of Concrete
## M-47, 08-96

STANDARD SPECIFICATIONS

FOR

REPAIR OF CONCRETE

August 1996

United States
Department of the Interior
Bureau of Reclamation
Technical Service Center
Denver Federal Center
Denver, Colorado 80225-0007

# Contents

HOW TO USE THIS MANUAL

This manual is divided into 3 sections.  The general requirements contained in
section 1 and the concrete preparation requirements contained in section 2
always apply to each and every Reclamation concrete repair project.  Section 3
consists of a number of paragraphs containing the special requirements of the
various repair materials and techniques.  Once a repair material/technique has
been properly selected from section 3 (see "Step 4. Choose An Appropriate
System" below) it should be sufficient in preparing a job specific
specification to simply require that the repairs be performed in accordance
with the provisions of section 1, section 2, and section 3.X of this manual.
Assistance in using this manual or with any other phase of concrete repair of
Reclamation structures can be obtained by contacting:

> Glenn Smoak or Kurt von Fay
> Materials Engineering and Research Laboratory
> Code D-8180
> Technical Service Center
> P.O. Box 25007
> Denver, Co  80225
> phone 303-236-3730.

The standard concrete repair materials and techniques described in this manual
have been developed and evaluated by the Bureau of Reclamation over a period
of some 90 years.  Unfortunately, during this time many repair failures have
occurred on Reclamation concrete structures even though very durable repair
materials were specified and used.  In evaluating the causes of these failures
it has been learned that it is essential to consistently use a systematic
approach to repairing concrete.  Many such such repair systems exist and this
manual will not attempt to discuss or evaluate which of these systems is best
for any set of field conditions.  Rather, the following repair system has been
used by Reclamation over a long period of time and has been found to result in
successful concrete repairs.  This system, known as "The Seven Steps of
Concrete Repair", is suitable for repairing both construction defects in new
concrete as well as old concrete damaged by long exposure to field conditions.
In using this system it is necessary to take the steps in numerical order.
Quite often the first questions asked when damaged concrete is detected is
"What should be used to repair this?" or "How much will the repair cost?".
These are not wrong questions.  Rather they are asked at the wrong time.  With
a systematic approach these questions are asked only when sufficient
information is available to provide correct answers.

The Seven Steps of Concrete Repair

1.   Determine the cause of damage

2.   Evaluate the extent of damage

3. Determine the need to repair

4. Choose an appropriate repair system

5. Prepare the old concrete

6. Apply the repair system

7. Cure the repair properly

Step 1. Determine the cause of damage - It is essential to correctly determine the causes(s) of the damage to the concrete. If this is not done, or if the determination is incorrect, the cause of damage will most likely attack and deteriorate the repair. The money spent for such repairs is, thus, totally lost and larger replacement repairs become necessary at much higher cost.

Step 2. Evaluate the extent of damage - The objective of this step is to determine how much of the structure is damaged and how extensive the damage is.

Step 3. Determine the need to repair - Not all damage to concrete requires repair. Repairs should be undertaken only if they will result in longer or more economical service life, a safer structure, or necessary cosmetic improvements in the structure. This step also includes determination of when the structure can be taken out of service for repairs, an estimate of how long the repairs will take, and how to budget the costs of the repairs.

These first 3 steps are the major components of a condition survey. Only after they have properly been performed should one proceed with selecting and installing the repair materials.

Step 4. Choose an appropriate repair system - Upon completion of the first 3 steps, an appropriate repair system can be selected that takes into consideration the many factors essential to a successful repair. In a majority of the repair situations, the standard materials and methods in this specification will fully meet all repair needs. It should be recognized, however, that service conditions or special repair situations will occur where these "standard materials" will not meet the special requirements and it will be necessary to resort to "non-standard" repair materials or methods. Such materials are continually being tested and evaluated by Reclamation laboratory and field offices. They include materials that may have performed quite well in laboratory tests but have not yet been applied in the field, materials not yet having long or sufficient field service to determine service life expectancy, or newly developed commercial materials not yet tested or evaluated by Reclamation. It is appropriate to use such repair materials and techniques only when it has been determined that none of the standard repair materials will properly serve, and when it is fully understood and agreed by all involved parties that the risks associated with the use of unproven materials are justified by the expected benefits of repair success and are acceptable.

Step 5. Prepare the old concrete - The most common cause of repair failure is improper or inadequate preparation of the old concrete prior to application of the repair material. Even the best of repair materials will give poor service life if bonded to weakened or deteriorated old concrete. The provisions of section 2 of this manual provide only the minimum preparation requirements. It should be noted that each of the standard repair materials has special preparation requirements and that these requirements are listed under paragraphs f of section 3. That is, for example, the special preparation requirements for replacement concrete will be listed under paragraphs 3.6.f while those of epoxy bonded concrete will be listed under paragraphs 3.8.f.

Step 6. Apply the repair system - Each standard and non-standard repair material has application procedures specific for that material. For example, the procedures used with replacement concrete are vastly different from those necessary for polymer concrete or epoxy bonded concrete. It is essential that proper application techniques as listed in the section 3 paragraphs g for each material be followed exactly.

Step 7. Cure the repair properly - The second most common cause of repair failures is improper or inadequate curing. Each repair material has specific curing requirements. As an example, replacement concrete benefits from long periods of water curing while latex modified concrete, currently a non-standard material, requires 24 hours water curing followed by drying to allow formation of the latex film. Polymer concrete has essentially no curing requirements while those of silica fume concrete are are exceedingly exact. Failure to achieve proper cure for the proper duration, as listed in section 3 paragraphs f, will result in loss of the repair at the final step of the process.

SECTION 1 - GENERAL REQUIREMENTS

1.1    GENERAL

Concrete that is damaged from any cause; structural concrete that has cracked; concrete that is honeycombed, fractured, or otherwise defective; and concrete that, because of excessive surface depressions, must be excavated and built up to bring the surfaces to the prescribed lines, shall be repaired or replaced in accordance with these specifications and contract requirements.

The method of repair or replacement (procedure) shall be as determined and directed by the Contracting Officer.

Contract, as used herein, shall mean the contract which, by reference, these specifications are included in and made a part.  Contracting Officer, as used herein, shall mean the person executing the contract on behalf of the Government, and shall include duly authorized representatives.  Contractor, as used herein, shall mean the party entering into the contract with the Government, and shall include subcontractors, suppliers, manufacturers, and agents at all tiers.

1.2    DESIGNS

Concrete mixture proportions, compressive strengths, and other design requirements shall be as specified in these specifications.  Designs by the Contractor shall be subject to the approval of the Contracting Officer.

1.3    SUBMITTALS

Submittals shall be in accordance with these specifications and the contract requirements entitled "Submittal Requirements."  The Contractor shall submit MSDS (Material Safety Data Sheets), approval drawings and data, repair plans, certifications, material samples, test data and samples, and other submittals as required in these specifications.  The Contractor shall be responsible for the accuracy of all submittals.

Unless specified otherwise submittals shall include the original and 2 copies.

1.4    QUALITY ASSURANCE

Quality assurance shall be in accordance with the requirements of the contract entitled "Quality Assurance," and with the requirements of these specifications.

All concrete shall be repaired as necessary to produce surfaces conforming to the specified tolerances and finish requirements of the contract in which these specifications are incorporated by reference and as outlined in section 3.

If, in the opinion of the Contracting Officer, the results of concrete repair indicate that proper quality control procedures are not being consistently utilized, further repair work may be suspended in whole or in part at the discretion of the Contracting Officer. Such suspension will be effective until the Contractor demonstrates substantial improvement in quality control procedures and repair results.

a. Contractor's quality control. - In accordance with the clause in the contract entitled "Inspection of Construction," the Contractor shall be responsible for providing quality control measures to ensure compliance of the repair with these specifications. The Contractor shall implement necessary and appropriate quality control procedures to ensure that all concrete repairs conform to the requirements of these specifications.

b. Government inspection and tests. - Government inspection and tests will be in accordance with the clause in the contract entitled "Inspection of Construction," and with these specifications.

Not less than 24 hours in advance of any concrete repair, the Contractor shall inform the Contracting Officer when the concrete repairs will be performed and, unless inspection is specifically waived, the repairs shall be performed only in the presence of an authorized representative of the Contracting Officer.

c. Testing. - Except as specified in paragraphs 3.5 and 3.12 of the specifications the Contractor shall perform all tests. The Contractor shall provide all materials, equipment, and labor necessary for performing the tests at no additional cost to the Government.

d. Approval. - Approval of the repairs will be based on the inspection and testing requirements of these specifications.

e. Rejection. - Repairs that fail to meet the requirements of these specifications will be rejected.

1.5   MATERIALS AND WORKMANSHIP

a. Materials. - The Contractor shall furnish all materials for repair or maintenance of concrete and shall furnish all materials for forming, curing, and protection of the repairs, as required. All materials shall meet materials specifications as specified in section 3, and all equipment used and methods of operation for the repair or maintenance of concrete shall be subject to approval of the Contracting Officer.

The references to materials in the specifications, wherein manufacturer's products or brands are specified by "brand name or equal" purchase descriptions, are made as standards of comparison only as to type, design, character, or quality of the article required, and do not restrict the Contractor to the manufacturer's products or to the specific brands named. It shall be the responsibility of the Contractor to prove equality of materials and products to those referenced and to provide all descriptive information, test results, and other evidence as may be necessary to prove the equality of materials or products which the Contractor offers as being equal to those referenced.

b. Workmanship. - Concrete shall be repaired by skilled workmen as outlined in section 3.

1.6    SAFETY

All work shall be performed in accordance with the applicable safety and health standards, the requirements of Reclamation's "Health and Safety Standards," the contract, and these specifications.  Certain additional safety precautions shall be employed to prevent skin and eye contact with chemicals, resins, or monomeric materials.  Protective  glasses and clothing, including rubber or plastic gloves shall be worn by all persons handling monomeric materials.  All exposed skin areas shall be protected with a protective barrier cream formulated for that purpose.  Barrier cream for skin protection shall be specified for the materials used and approved by a physician. Adequate ventilation shall be provided and maintained at all times during use of monomeric materials and solvents.  Fans used for ventilating shall be explosion proof.  If necessary, respirators that filter organic fumes and mists shall be worn.  All contaminated materials such as wipes, empty containers, and waste material shall be continually deposited in containers that are protected from spillage.  Spillage shall be immediately and thoroughly cleaned up and disposed of in accordance with applicable regulations.

Federal Standard No. 313, as amended, for the preparation and submission of material safety data sheets is hereby incorporated and made a part of these specifications.

In accordance with the clause in the contract entitled "Hazardous Material Identification and Material Safety Data," the Contractor shall submit a completed MSDS (Material Safety Data Sheet), Department of Labor Form OSHA-174, or GSA-approved Alternate Form A for each hazardous material as required by Federal Standard No. 313, as amended.  The information in this MSDS shall be followed to assure safe use, handling, storage, and an environmentally acceptable disposal of the commodity used on the job site.

The Contractor shall submit to the Contracting Officer, not less than 30 days prior to job site delivery of each hazardous material, completed MSDS and identification and certification for the material.

1.7    REPAIR PROCEDURES

All repair or maintenance procedures shall be performed in accordance with the applicable specifications of section 3, including surface preparation, forming, finishing, and curing.

# SECTION 2 - CONCRETE PREPARATION FOR REPAIR

Concrete to be repaired shall be prepared in accordance with the requirements of this paragraph and with the special requirements of section 3.

## 2.1    REMOVAL AND CLEANING

All damaged, deteriorated, loosened, or unbonded portions of existing concrete shall first be removed by water blasting, bush hammering, jack hammering, or any other approved method, with approved equipment, after which the surfaces of existing concrete shall be prepared by contained shotblasting, wet sandblasting, or water blasting to remove any microfractured surfaces resulting from the initial removal process. The surfaces shall then be cleaned and allowed to dry thoroughly, unless the specific repair technique requires application of materials to a saturated surface. Concrete removal processes involving the use of jack hammers in excess of 30 pounds, dry sandblasting, or scrabblers shall not be used without approval by the Contracting Officer. The use of acids for cleaning or preparing concrete surfaces for repair will not be permitted.

## 2.2    SAW CUT EDGES

The perimeters of repairs to concrete that involve concrete removal and subsequent materials replacement shall be saw cut perpendicular to the repair surface to a minimum depth of 1 inch. Featheredge repairs to concrete shall not be used.

## 2.3    REINFORCING STEEL

All loose scale, rust, corrosion by products, or concrete shall be removed from exposed reinforcing steel. Reinforcing steel exposed for more than one-third of its perimeter circumference shall be completely exposed to provide 1-inch minimum clearance between the steel and the concrete. Damaged or deteriorated reinforcing steel shall be removed and replaced as directed by the Contracting Officer.

## 2.4    MAINTENANCE OF PREPARED SURFACES

After the concrete has been prepared and cleaned, it shall be kept in a clean, dry condition until the repair has been completed. Any contamination, including oil, solvent, dirt accumulation, or foreign material shall be removed by additional wet sandblasting and air-water jet cleanup followed by drying.

# SECTION 3 - SPECIFICATIONS FOR REPAIRING CONCRETE

## 3.1    SURFACE GRINDING

Where bulges, offsets, and other irregularities exceed the specified tolerances, the protrusions shall be repaired so that the surfaces are within the specified limits. Surface grinding techniques may be used for this purpose subject to the following limitations:

   a. Grinding of surfaces subject to cavitation erosion (hydraulic surfaces subject to flow velocities exceeding 40 feet per second) shall be limited in depth so that no aggregate particles are exposed more than 1/16 inch in cross section at the finished surface.

   b. Grinding of surfaces exposed to public view shall be limited in depth so that no aggregate particles are exposed more than 1/4 inch in cross section at the finished surface.

   c. Grinding of all other surfaces shall be as directed by the Contracting Officer. In no event shall surface grinding result in exposure of more than one-half the diameter of the maximum-size aggregate.

   d. Where surface grinding has caused or will cause exposure of aggregate particles greater than the limits of subparagraph 3.1.a. or b., the concrete shall be repaired by excavating and replacing the concrete in accordance with paragraph 3.6 or 3.8. as directed by the Contracting Officer.

3.2   PORTLAND CEMENT MORTAR

   a. General. - Repairs with portland cement mortar shall be made only if specifically approved by the Contracting Officer. Approval for hand-applied cement mortar repairs will be given only for very small repair areas not associated with critical performance of the structure. When approved, portland cement mortar may be used for repairing defects on exposed, new concrete surfaces if the defects are small and are too wide for dry pack and too shallow for concrete replacement and only if the repairs can be completed within 24 hours of removing the forms. Portland cement mortar shall not be used for repairs to old or existing concrete or for repairs that extend to or below the first layer of reinforcing steel.

   b. Submittals. - Before beginning any repair work, the Contractor shall submit a detailed list of the equipment, procedures, and materials the Contractor proposes to use for cement mortar repair to the Contracting Officer for approval.

   c. Quality assurance. - Quality assurance shall be in accordance with paragraph 1.4.

   d. Materials. - Portland cement mortar shall consist of type I or type II portland cement, clean water, and clean, well-graded sand passing a 1.18-mm (No. 16) sieve. All mortar materials, including curing compound shall meet the requirements of subparagraph 3.6.d.

   e. Safety. - All work shall be performed in accordance with the requirements of paragraph 1.6.

f.  Concrete preparation. - After damaged or defective concrete has been removed as specified in section 2, the surface on which mortar is to be placed shall be prepared by being thoroughly cleaned of all micro fractured, loose, or deteriorated materials and surface-dried.

g.  Application and Mixture Proportioning. - Portland cement mortar shall be composed of portland cement, sand, and water, all well mixed and brought to proper consistency.  Mortar mixtures and application techniques shall be in accordance with the requirements for mortar replacement method as described in Reclamation's "Concrete Manual", Eighth Edition, revised, chapter VII.  Cement mortar shall not be applied until approval of all submittals has been received from the Contracting Officer.

h.  Curing. - All cement mortar repairs shall be water cured for 7 days following application.  At no time during this initial curing period shall the mortar be allowed to dry.  If drying occurs, the repair shall be removed and replaced.  If the repair is removed and the original concrete is older than the maximum specified in subparagraph 3.2.a., another repair method shall be used in accordance with these specifications.  Following the 7-day curing period and while the repair is still saturated, the surface of the repair shall receive two coats of wax-base (type I) or water-emulsified resin base (type II) curing compound meeting the requirements of subparagraph 3.6.d.

## 3.3    DRY PACK AND EPOXY BONDED DRY PACK

a.  General. - The dry pack concrete repair technique shall be limited to areas that are small in width and relatively deep, such as core holes, holes left by the removal of form ties, cone-bolt and she-bolt holes, and narrow slots cut for repair of cracks.  Epoxy bonded dry pack shall be used for critical repairs or for repairs expected to be exposed to severe service conditions.  Dry pack shall not be used for shallow depressions where lateral restraint cannot be obtained, nor for filling behind steel reinforcement.

b.  Submittals. - Before beginning any repair work, the Contractor shall submit to the Contracting Officer for approval, data for the mortar materials and as provided by paragraph 3.8.b.(2).

c.  Quality assurance. - Quality assurance shall be in accordance with paragraph 1.4.

d.  Materials. - Dry pack mortar shall consist of type I or II portland cement, clean sand that will pass a 1.18-mm (No. 16) sieve, and clean water.  All dry pack materials, including curing compound, shall meet the requirements of subparagraph 3.6.d.  Epoxy bonding resin shall meet the requirements of paragraph 3.8.d.(2).

e.  Safety. - All work shall be performed in accordance with the requirements of paragraph 1.6.

f. Concrete preparation. - Holes for dry pack shall have a minimum depth of 1 inch and shall be square at the surface edge. A careful inspection shall be made to ensure that the hole is thoroughly clean and is in sound concrete. The interior surfaces of the hole shall be presoaked prior to application of the dry pack. If epoxy bonded dry pack is to be used presoaking the repair area prior to application of the dry pack shall not be performed.

g. Application. - A mortar bond coat or epoxy resin bond coat shall be applied to the concrete hole surface prior to placing dry pack. The mortar bond coat shall consist of 1 part portland cement to 1 part sand mixed with water to give a fluid paste consistency. The mortar bond coat shall be thoroughly brushed onto the hole surfaces. The epoxy resin bond coat shall consist of materials and be applied in accordance with the provisions of paragraph 3.8.g.(3). Dry pack mortar shall consist of 1 part portland cement to 2.5 parts sand (by weight) and shall be mixed with just enough water so that the mortar will stick together when molded by hand and not exude water when squeezed.

The dry-pack mortar shall be immediately packed into place in 3/8-inch compacted layers before the bond coat has dried or cured.

Each layer shall be compacted over the entire surface by tamping with a hardwood dowel and hammer. Hardwood dowels are used in preference to metal bars because the bars tend to polish the surface of each layer. The final layer shall be finished immediately after compaction by laying the flat side of a hardwood board against the fill and striking the board firmly with a hammer.

h. Curing. - Proper curing is essential for a successful dry-pack repair. The surface of the repair area shall be protected from drying and shall be kept continuously moist for 7 days. The surface of the repair shall be protected from surface drying by using burlap kept wet, wet sand, or plastic sheeting over a water soaker hose, or other methods approved by the Contracting Officer. After 7 days and while the surface is still damp, two coats of curing compound shall be applied to prevent moisture loss. The dry-pack repair area shall not be exposed to freezing temperatures for 3 days after application of the curing compound.

3.4   PREPLACED AGGREGATE CONCRETE

a. General. -

(1) Preplaced aggregate concrete (PAC) is concrete that has been made by forcing a grout into the voids of a mass of clean, graded, coarse aggregate. As a repair method, PAC is used where placing conventional concrete is extremely difficult, such as in underwater repairs, concrete and masonry repairs, or where shrinkage of concrete must be kept to a minimum. In underwater repairs, injection of grout at the bottom of the

PAC displaces water, leaving a homogeneous mass of concrete with a minimum of paste washout.

For the purpose of this repair method, grout is defined as a mixture of water, cementitious materials, sand, and admixtures.

(2)  Reference standards and specifications for PAC. - Reference standards and specifications for materials, testing, and proportioning PAC shall be the standards listed below.  Standards and specifications for testing concrete and concrete materials shall be in accordance with applicable ASTM and ACI standards.

(a)  ASTM C 937 - Standard Specification for Grout Fluidifier for Preplaced-Aggregate Concrete.

(b)  ASTM C 938 - Standard Practice for Proportioning Grout Mixtures for Preplaced-Aggregate Concrete.

(c)  ASTM C 939 - Standard Test Method for Flow of Grout for Preplaced-Aggregate Concrete (Flow Cone Method).

(d)  ASTM C 940 - Standard Test Method for Expansion and Bleeding of Freshly Mixed Grouts for Preplaced-Aggregate Concrete in the Laboratory.

(e)  ASTM C 941 - Standard Test Method for Water Retentivity of Grout Mixtures for Preplaced-Aggregate Concrete in the Laboratory.

(f)  ASTM C 942 - Standard Test Method for Compressive Strength of Grouts for Preplaced-Aggregate Concrete in the Laboratory.

(g)  ASTM C 943 - Standard Practice for Making Test Cylinders and Prisms for Determining Strength and Density of Preplaced-Aggregate Concrete in the Laboratory.

(h)  ASTM C 953 - Standard Test Method for Time of Setting of Grouts for Preplaced-Aggregate Concrete in the Laboratory.

(i)  American Concrete Institute Standard (ACI-304), Recommended Practice for Measuring, Mixing, Transporting, and Placing Concrete.

b.  Submittals. -

(1)  At least 60 days prior to beginning repair work, the Contractor shall submit for approval a repair plan for PAC construction.  The repair plan shall include detailed drawings of formwork construction, the grout injection system including sequence of injecting grout into insert pipes, location and spacing of injection tubes, sounding wells, and pipes; a description of all equipment including pumps, sizes and diameters of pipes, hoses, and connections; the planned operating

procedures and pumping rates for grout; methods of placing coarse aggregate; methods of consolidating aggregates and PAC during grouting; and communication facilities between the grout mixing and pumping plant and PAC placement.

The proportions of grout mixtures and fresh properties of grout shall be submitted with the repair plan. Included with the grout proportions, shall be the voids content and compacted density of aggregates before injection of grout, the density of hardened PAC, and the compressive strengths of PAC at 7 and 28 days' age.

(2) The Contractor shall submit for approval to the Contracting Officer's representative a proposed specific operating procedure including a hazard analysis for all preplaced aggregate concreting and grout injecting operations. The specific operating procedure shall include such things as engineering controls, protective clothing, eye protection, respiratory protection, and air sampling as necessary to check the effectiveness of the control program.

(3) Pump rating curves and complete mixer details, including photographs or drawings of the proposed mixing equipment, shall be submitted to the Contracting Officer for approval 30 days prior to use. The Contracting Officer shall have the right to require the Contractor to make changes in the equipment which the Contracting Officer determines necessary to make the equipment perform satisfactorily during the grouting operations without additional cost to the Government

(4) The Contractor shall submit test results from testing of cylinders in accordance with the requirements of subparagraph c. below.

c. Quality assurance. - Quality assurance shall be in accordance with paragraph 1.4 and these specifications. Unless otherwise directed, grout mixtures shall be proportioned in accordance with ASTM C 938.

For normal structural work, the ratio of cementitious materials (cement plus pozzolan) to sand shall be 1:1. If leaner mixes are required (generally for low heat generation), the ratio may be adjusted, but shall not exceed 1:2.

The ratio of cement to pozzolan, by weight, shall be between 2:1 and 4:1 as directed by the Contracting Officer.

The ratio of water to cementitious materials

$$\frac{W}{P+C}$$

where:
  W = weight of water
  P = weight of pozzolan
  C = weight of concrete

M-47 (M0470000.896)
Page 13 of 73
8-1-96

shall not exceed 0.50 by weight. For severe exposure conditions, the W/P+C ratio shall be in accordance with Reclamation's "Concrete Manual," Eighth Edition, revised, chapter III, table 15.

The pumpability of grout containing fine sand (grading 1, table 2 of ASTM C 637) shall be controlled by the consistency test using the flow cone in accordance with ASTM C 939. The flow time shall be between 20 and 30 seconds unless it can be demonstrated that grout can be effectively pumped at a different flow time.

Compressive strength cylinders shall be cast in accordance with ASTM C 943 and tested at 7 and 28 days' age with a minimum of two cylinders tested for each age. The design compressive strength of PAC shall be 4,000 pounds per square inch at 28 days' age. The required average strength for PAC is that which will ensure that 80 percent of all test cylinders exceed the design strength.

The quality of the top of formed PAC placements shall be ensured by venting air, water, and low-quality grout from the uppermost location in the placement.

For unformed surfaces where a screeded or troweled finish is required, the grout level shall flood the top surface above any aggregate, and any diluted grout shall be removed by brooming. A thin layer of pea gravel having a maximum-size aggregate (MSA) of 3/8 inch shall be worked into the surface by raking and tamping or internal vibration. The surface shall be finished in accordance with conventional concrete procedures to resemble the finish of the surrounding concrete. Care shall be taken when topping off the unformed surface to avoid lifting or loosening of the surface aggregate.

d. Materials. - Materials for PAC shall be in accordance with the requirements of ASTM C 938 and the following additional requirements:

   (1) Pozzolan. - Pozzolan shall meet the requirements of ASTM C 618 for class F pozzolan.

   (2) Admixtures. -

      (a) Air-entraining admixture. - Air-entraining admixture (AEA) shall meet the requirements of ASTM C 260.

      The amount of AEA used for injecting mortar shall be that amount necessary to effect a total air content in the mortar, prior to injection, of 9 percent, plus or minus 1 percent, by volume of mortar.

      (b) Grout fluidifier. - Grout fluidifier shall meet the requirements of ASTM C 937.

(c) Chemical admixtures. - Chemical admixtures, if permitted, shall meet the requirements of ASTM C 494 for types A, D, F, or G admixtures. Chemical admixtures shall not be used without prior approval from the Contracting Officer.

(3) Sand. - Sand shall consist of natural particles which may be supplemented by limited quantities of crushed sand to make up for deficiencies in the natural materials. Natural sand is required due to its favorable particle shape for pumping grout. Sand for PAC differs from sand used for conventional concrete, primarily in its finer grading.

Sand shall meet the quality requirements of subparagraph 3.6.d.; and the grading requirements of ASTM C 938.

(4) Coarse aggregate . - Coarse aggregate shall be clean and well-graded. Unless otherwise directed, the MSA shall be 1-1/2 inches, and after compaction in the forms, shall have a voids content from 35 to 45 percent.

Coarse aggregate shall meet the quality requirements of subparagraph 3.6.d.; and, the grading requirements of ASTM C 637, table 2, grading 1 for coarse aggregate.

For larger placements, the largest size aggregate that can economically be placed shall be used. The MSA and grading of coarse aggregate for these placements should follow the guidelines of ACI 304.

e. Safety. - All work shall be performed in accordance with requirements of paragraph 1.6 and these specifications.

The Contractor shall comply with Reclamation's "Construction Safety Standards" as well as all other applicable safety and health standards and regulations. In the event of conflict, the more stringent standard shall apply.

f. Concrete preparation. - The Contractor shall prepare all concrete surfaces and foundations for PAC application. Concrete surface preparation shall be in accordance with subparagraph 3.6.f.

For underwater repairs, high-pressure water jetting shall be the normal method for concrete removal and surface scarification. Low-pressure water jetting will be allowed for cleanup immediately before placing PAC.

In underwater construction where contamination is known or suspected to exist, the water shall be sampled and tested to determine the degree of contamination and its possible influence on the quality of concrete. Where moderate contamination is present, concrete surfaces shall be cleaned within 2 days prior to placing aggregate and grouting. If contaminants are present in such quantity or are of such character that harmful effects cannot be eliminated or controlled, preplaced aggregate concrete shall not be used.

All loose, fine material shall be removed from the placement insofar as possible before placement of coarse aggregate to prevent subsequent coating of the aggregate or filling of voids.

g.  Application. -

    (1)  General. - Application of PAC shall include construction and placement of formwork and the grout injection system, placement of aggregate, injection of grout, finishing, and curing of PAC.

    (2)  Equipment. - The grout mixing, pumping, and injection system shall be furnished by the Contractor.

The grout injection system shall be designed to deliver and inject grout into the preplaced aggregate, to provide a means for determining the grout elevations within the aggregate mass, and to provide vents in enclosed forms for water and air to escape.  The injection system shall have a bypass for returning grout to an agitating tank.

Pumps shall be of a positive displacement type and be equipped with a pressure gauge on the outlet line to indicate any incipient line blockage.  At least one extra pump shall be on standby to maintain continuous pumping operations.  The pumps shall have quick-disconnecting fittings to the grout supply line.

Mixing and agitating equipment shall be sized to maintain a continuous, uninterrupted flow of grouting mortar for the duration of the PAC placement.  The mixing plant shall be a high-speed centrifugal type and shall be equipped with an accurate water meter, reading cubic feet to tenths of a cubic foot.  In addition to the grout mixer, a holdover mechanical agitator tank of similar volume as the mixer shall be provided.  Suitable provisions shall be made for passing the grout through a U.S. Standard 2.36-mm (No. 8) sieve as it is discharged from the mixer.

The batching and mixing plant equipment shall accurately measure the amounts of cement, pozzolan, sand, admixtures, and water batched for grout.  Inaccuracies in feeding and measuring during operation shall not exceed by individual weight plus or minus 1 percent for water; plus or minus 2 percent for cement, pozzolan, or sand; and plus or minus 3 percent for admixtures.

The length of delivery line shall be kept to a practicable minimum. Pipe sizes shall be designed so that during operation, the grout velocity ranges between 2 and 4 feet per second for delivery lines up to 300 feet, or at a pumping rate of about 1 cubic foot of grout per minute through a 1-inch-diameter pipe.  Pipe sizes shall be approximately

1 inch in diameter, but may need to be increased in size for delivery lines longer than 300 feet to reduce pressures.

Grout insert pipes shall have a diameter of 3/4 to 1 inch and shall be placed vertically, horizontally, or at angles to inject grout at the proper location. Grout insert pipes shall be in sections about 5-1/2 feet in length for ease of withdrawal. For depths greater than 15 feet, grout insert pipes shall be flushed coupled. For depths less than 15 feet, standard pipe couplings may be used.

Connections between grout delivery lines and insert pipes shall have quick-disconnect fittings. Quick-disconnect fittings which reduce the cross section of the flow area shall not be used. Each insert pipe, or the end of the delivery line where it attaches to the insert pipe, shall be equipped with an individual valve to control or stop the flow of grout. Valves shall be of a quick-opening, plug type which can readily be cleaned.

For vertical and angled injection tube placements, sounding wells shall be used to determine the level of grout in the placement. Sounding wells shall consist of slotted pipe with a diameter of about 2 inches. A 1-inch-diameter float, weighted so that it will float on grout and sink in water (for underwater placements), shall be attached to a sounding line and float freely in the sounding well.

(3) Formwork. - Forms for PAC may be of wood, steel, or other materials suitable for conventional concrete. Wood forms or other materials which may swell or become damaged by submergence in water shall not be used for underwater repairs.

Joints and any entry holes for bolts, reinforcement, injection ports, vents, or injection wells shall be caulked to eliminate grout leakage.

Forms shall be designed to resist the lateral pressure exerted by aggregates during and after placing, and to resist full hydraulic pressure of the concrete throughout placing until initial setting has begun.

Form vibrators for consolidation shall be mounted so that vibration does not loosen bolts and joints which may lead to grout leakage or failure. Form vibrators shall be mounted on ribs or stiffeners attached to the sheathing to effectively transmit vibration to the concrete.

All forms shall be coated with a suitable bond breaker for ease of removal without damaging concrete. For underwater repairs, the bond breaker shall be selected so that it does not wash off the forms during submergence. Formwork, injection tubing, inspection wells, and other equipment shall be protected from damage by aggregates during placing.

Reinforcing steel shall be installed during construction of formwork and securely held in place during placement of aggregate and injection of grout.

(4) Grout pipe system. - The grout pipe system shall be furnished by the Contractor and shall consist of a series of regularly spaced injection tubes with outlets beginning at the bottom and continuing to the top of the compacted aggregates. Observation wells shall be spaced at regular intervals between the injection tubes. Vent pipes shall be used in forms that contain restricted or irregular spaces where water or air may be entrapped by the rising grout surface, such as blockouts or embedment work.

The grout pipe system shall be designed to inject grout beginning at the lowest point in the PAC placement and continue to raise the grout level within the mass by selectively withdrawing the pipes and switching to other pipes to maintain the grout at a gently sloping to nearly horizontal level. The grout pipe system shall be coded to clearly identify the location of the outlet of the pipe supplying grout and the order of supplying grout to the pipes.

The spacing of grout insert pipes shall be between 4 and 12 feet and should average approximately 6 feet. The outlet of grout insert pipes shall begin not more than 6 inches above the bottom of the aggregate mass. Rows of horizontal pipes should be not more than 6 inches above the next lower row of pipes.

As a guide for layout of insert pipes, it can be assumed that the grout surface will have a 1:4 (vertical to horizontal) slope in the dry and a 1:6 slope under water.

For vertical and inclined injection systems, sounding wells shall be spaced to observe the level of grout. The ratio of number of sounding wells to insert pipes shall be between 1:4 and not less than 1:10. Sounding wells shall be placed vertically or near vertically.

The delivery line shall consist of a single pipeline extending from the pump to the insert pipe equipped with a valve, or to a wye branching to two insert pipes equipped with valves at the beginning of the wye and at both outlets. A manifold system in which more than one grout insert is operative at the same time shall not be used.

(5) Aggregate preparation and placement. - Coarse aggregate shall be washed and screened immediately before placing in the forms. If more than one size of coarse aggregate is used, the aggregate shall be batched and mixed in the proper proportions or discharged at proportional rates onto a vibrating deck or revolving wash screens.

Coarse aggregate shall be transported to the forms by buckets, conveyors, or other approved methods. Coarse aggregates shall be dumped as close as possible to their final location. A flexible rubber elephant trunk (tremie) shall be used to limit free fall to less than 5 feet to minimize segregation and breakage of aggregate. For large placements, a gated pipe with a diameter at least four times the MSA shall be used. The pipe shall be gradually filled and lowered toward the point of placement, and after the pipe is completely filled at the upper end, aggregate shall be discharged from the lower gate. Aggregate shall not be discharged through an unfilled or partially filled pipe.

Coarse aggregate shall be dumped and spread in nearly horizontal lifts, each lift not to exceed 1 foot in height. Around closely spaced embedded items, such as reinforcing steel, conduits, and blockouts, and in more difficult placements where high density and exceptional homogeneity of concrete are desired, the lift height shall be limited to no more than 4 inches and may have to be placed by hand. Vehicular traffic shall not be permitted on top of preplaced coarse aggregate, with the exception of small skips or loaders and only with approval of the Contracting Officer. This equipment shall have rubber tires and be cleaned of any loose debris or substances that may contaminate aggregates or interfere with grout travel during injection.

Although coarse aggregate shall be completely washed prior to placing it shall not be flushed with water in the forms for the purpose of washing.

If water is used for the purpose of precooling aggregate or to provide lubrication for grout, the water shall be injected through the preplaced grout insert pipes rather than on top of the aggregates.

(6) Grout injection. - Injection shall begin at the lowest insert pipe point within the form and continue to raise the level of grout evenly to the top of the mass of aggregate. The injection process shall raise the level of grout approximately 18 to 24 inches above the outlet of the insert pipe before withdrawing of that pipe begins. Thereafter, the level of grout shall remain a minimum of 12 inches above the outlet of the insert pipe until the injection is completed.

Grouting shall proceed at only one insert pipe at a time. If a wye or manifold system is used, the system shall be equipped with individual valves to control the flow to each insert pipe.

Before the insert pipe is withdrawn, the valve at the pipe inlet shall be shut off. The delivery pipe may then be disconnected to inject at other locations.

When injection insert pipes are placed horizontally, the process shall begin at the lowest insert pipe. Adjacent insert pipes shall have valves opened both alongside and above the pipe being injected. When grout begins to flow out of these pipes, they shall be shut off and grouting continued so that the adjacent pipes are completely submerged.

The initial pipe shall then be shut and the process shall be repeated, traveling in a horizontal direction before proceeding to the upper row.

Grout shall not be allowed to set up in the insert pipes between injections. If the injection process is expected to continue for an extended period, a suitable retarding admixture shall be used in the grout. Vertical insert pipes may be rodded to remove grout prior to reinjection; however, insert pipes shall not be cleaned by injecting water through them.

The rate of grout rise and rate of grout injection shall be controlled within the form so that excessive form pressures are not created and so that grout does not cascade within the mass. This is particularly important when grouting under water to avoid sand streaks and honeycombing within the mass.

When grouting around embedded items and particularly under flat surfaces or recessed areas, the grout shall be injected until good quality grout is forced from vent pipes. Good quality grout shall be grout that is not diluted and free from sand, silt, trash, and other items. The rate of flow of grout and grout injection pressures shall be closely monitored in enclosed locations to avoid form failure.

Form vibration shall be used to consolidate grout and remove entrapped air pockets adjacent to form surfaces. Excessive form vibration which causes sand streaking shall not be done. Internal vibration shall not be used for consolidation of the aggregate-grout mixture, but may be used for topping off unformed surfaces.

Unplanned construction joints shall not be used without approval from the Contracting Officer. If a breakdown occurs that stops placement, the grout insert tubes shall be withdrawn, and the grout surface shall be considered a cold joint. Cold joints shall be treated by first removing coarse aggregate down to the joint surface, removing aggregate and poor-quality mortar, and then cleaning the joint surface by approved methods, taking care not to undercut exposed aggregates.

h. Curing. - Curing and protection of PAC shall be in accordance with subparagraph 3.6.h.

3.5    SHOTCRETE

a. General. - Shotcrete is defined as pneumatically applied concrete or mortar placed directly onto a surface. The shotcrete shall be composed of water, cementitious materials, sand, coarse aggregate, steel fibers (if specified), and admixtures, and shall be placed by either the dry-mix or wet-mix process as specified herein.

The dry-mix process shall consist of thoroughly mixing the solid materials; feeding these materials into a mechanical feeder or gun; carrying the materials by compressed air through a hose to a special nozzle; introducing the water and intimately mixing it with the other ingredients at the nozzle; and then jetting the mixture from the nozzle at high velocity onto the surface to receive the shotcrete.

The wet-mix process shall consist of thoroughly mixing all the ingredients with the exception of the accelerating admixture, if used; feeding the mixture into the delivery equipment; delivering the mixture by positive displacement or compressed air to the nozzle; and then jetting the mixture from the nozzle at high velocity onto the surface to receive the shotcrete.

The equipment used by the Contractor for mixing and applying shotcrete shall be capable of handling and applying shotcrete containing the specified maximum-size coarse aggregate. All equipment, including mixers, hoses, nozzles, nozzle liners, air- and water-pressure gauges, and gaskets, shall be maintained in clean and proper operating condition satisfactory to the Contracting Officer.

b. Submittals. - The Contractor shall submit for approval to the Contracting Officer a proposed specific operating procedure including a hazard analysis for all shotcrete operations. The specific operating procedure shall include such things as engineering controls, protective clothing, eye protection, respiratory protection, and air sampling as necessary to check the effectiveness of the control program.

The Contractor shall submit the specimens extracted from panels fabricated for preapplication testing.

The Contractor shall submit test specimens of fresh and hardened concrete from locations directed by the Contracting Officer.

c. Quality assurance. - Quality assurance shall be in accordance with paragraph 1.4 and these specifications. The Government will perform all testing of fresh and hardened shotcrete. The Contractor shall obtain specimens from locations specified by the Government. The compressive strength of the shotcrete will be determined through the medium of tests of 3- by 3-inch cores or 3-inch cubes. The average compressive strength of specimens taken from a shotcrete application shall be not less than:

(1) Four thousand pounds per square inch at 28 days' age.

(2) Six hundred pounds per square inch at 8 hours' age. This will be determined from 3-inch cubes extracted from test panels.

Adjustments shall be made as directed by the Contracting Officer to obtain shotcrete having suitable impermeability, strength, density, and durability. Suitable strength is that which will ensure that 80 percent of all test specimens exceed or equal the average strength as specified above.

(3) Preapplication testing of shotcrete and nozzlemen. - No shotcrete shall be applied to the work until preapplication testing indicates that the nozzleman is qualified and the proposed shotcrete mixture meets the specified compressive strength. The Contractor shall allow adequate lead time for fabricating of test panels and testing shotcrete specimens. Furthermore, the Contractor is encouraged to fabricate panels for several mixtures since failure of a single mixture to meet specified strength requirements could delay accomplishment of other work.

(a) Shotcrete mixture. - At least 30 days prior to application of shotcrete to the work, the Contractor shall fabricate two test panels for each mixture to be used for vertical and overhead positions of application.

Application of shotcrete to test panels may be accomplished at locations other than the construction site provided:

(aa) Equipment used for fabricating test panels is identical to that to be used in application.

(bb) The materials used in fabricating test panels are from the same sources as those to be used in application and meet all specifications requirements.

(cc) Application is made in the presence of the Contracting Officer by a nozzleman who will later apply the shotcrete to the work.

Test panels shall be fabricated by applying shotcrete in one application not less than 4 inches thick to a panel form made of plywood or other suitable material.

The panel form to be used for vertical and overhead applications shall be either of two types as follows:

(dd) Open ended on at least two sides with 90° edges on enclosed sides. The minimum size of this open-ended panel shall be 18 inches square.

(ee) Enclosed on all four sides with members that taper outward in a manner that prevents the accumulation of rebound at edges and corners while allowing a shotcrete thickness of not less than 4 inches over the entire flat area of the panel. The minimum size of the center flat area of panels enclosed by tapered edges shall be 15 inches square.

Panels should be stiffened and weighted sufficiently to provide rigid surfaces against which the shotcrete can be applied. The panels shall contain the same type of reinforcement as to be used in construction to indicate whether sound shotcrete is obtained without shadowing behind the reinforcement.

After fabrication, the panel shall, except when test specimens are being obtained, be covered and sealed with polyethylene sheeting or shall be cured by any other approved method that will prevent loss of moisture.

The Contractor shall provide all necessary equipment and shall obtain 3-inch cubes from the test panels for testing the compressive strength at 8 hours' age and shall obtain 3-inch cubes or 3-inch-diameter cores from test panels for testing the compressive strength at 28 days' age.

All core drilling and saw cutting of cubes shall be performed in a workmanlike manner by competent and experienced workmen. Test specimens shall be extracted at the latest time, satisfactory to the Contracting Officer, that the specimens can be delivered by the Contractor to the Government for testing at the specified age. At least five specimens shall be obtained from each of the panels for each mixture being verified. Extreme care shall be taken in drilling cores or cutting cubes to obtain specimens satisfactory to the Contracting Officer.

Cores shall be 3 inches in diameter, drilled the full depth of the shotcrete; they shall be straight, sound, of uniform diameter, and of sufficient length for subsequent trimming by the Contractor to produce a right cylinder with an L/D ratio of 1.0 plus or minus 0.03.

Cubes shall be cut full depth of the shotcrete, shall be sound, and shall have six uniform square sides of 3-inch length plus or minus 0.05 inch.

Immediately after extraction, test specimens shall be individually wrapped and sealed in polyethylene bags or covered and sealed by any other approved means to prevent loss of moisture. Each test specimen shall be properly marked for identification.

Each set of test specimens shall be delivered without delay to the project laboratory, or other location approved by the Contracting Officer, for testing at the specified age.

(b)  Shotcrete nozzleman. - The nozzleman who applies shotcrete shall be certified with the materials and equipment to be used for the work. Certification shall be in accordance with ACI 506.3R with the following limitations:

   (aa)  The certification examination will be conducted by authorized representative of the Contracting Officer.

   (bb)  The examination will consist of an oral test and a field workmanship demonstration test.

(cc) The test panels shall be a minimum of 18 inches square and 4 inches deep. The sides may have beveled edge forms (45° angle out to reduce the entrapment of rebound in the corners).

(dd) Test specimens shall be of the size and shape specified for quality control in these specifications.

d.  Materials. -

(1)  Portland cement. - The cementitious materials in shotcrete shall comply with subparagraph 3.6.d.(1), with the exception that type III cement may be approved for use if sulfate conditions do not exist.

(2)  Water. - Water shall be in accordance with subparagraph 3.6.d.(4).

(3)  Sand and coarse aggregate. - Except as hereinafter provided for coarse aggregate grading, the sand and coarse aggregate shall be in accordance with subparagraph 3.6.d.(5).

The maximum-size coarse aggregate shall be no larger than 3/8 inch. No material retained on the 3/8-inch sieve shall be permitted. Only 3 percent significant undersize aggregate material that will pass a 4.75-mm (No. 4) U.S. Standard sieve will be permitted.

(4)  Admixtures. - Admixtures shall be in accordance with subparagraph 3.6.d.(3).

(a)  Accelerator. - The Contractor may use an accelerating admixture in shotcrete.

(b)  Air-entraining admixture. - The amount of air-entraining admixture used for the wet-mix process shall be that amount necessary to effect a total air content in the shotcrete, prior to application, of 7 percent plus or minus 1 percent by volume of shotcrete.

(5)  Steel fibers. - Where directed by the Contracting Officer, the Contractor shall furnish steel fibers for reinforcement. The amount of steel fibers used shall be 90 pounds of fiber per cubic yard of shotcrete.

Steel fibers shall be carbon steel deformed type I (cold drawn wire) or type II (cut sheet) to conform to the requirements of ASTM A 820. Length of steel fiber shall be 3/4-inch minimum to 1-1/4-inch maximum. The length divided by diameter (or equivalent diameter), or aspect ratio, shall be 45 minimum and 100 maximum. The steel fiber shall have a minimum average tensile strength of 50,000 pound-force per square inch.

(6)  Curing compounds. - Curing compounds shall meet the applicable requirements of subparagraph 3.6.d.(6).

e. Safety. - All work shall be performed in accordance with the requirements of paragraph 1.6.

f. Concrete preparation. - Concrete to be repaired with shotcrete shall be prepared in accordance with section 2.

g. Application. -

(1) Mixture proportions. - The proportions of water, cementitious materials, sand, coarse aggregate, and admixture and fibers, if used, shall be determined by the Contractor subject to approval by the Contracting Officer to obtain the specified compressive strength and bond. The shotcrete shall have a minimum cementitious materials content of 658 pounds per cubic yard, as discharged from the nozzle. Furthermore, the amount of cementitious materials will be increased by the Contracting Officer as necessary to obtain the specified compressive strengths and bond.

(2) Consistency. - The consistency of the dry-mix process shotcrete shall be regulated by the amount of water introduced at the nozzle and shall be adjusted so that the in-place shotcrete is adequately compacted and neither sags nor shows excessive rebound.

The consistency of the wet-mix process shotcrete at the delivery point shall not exceed a 3-inch slump.

(3) Batching. - Batching for the dry-mix process shall be as specified in this paragraph. Water, cementitious materials, sand, coarse aggregate, admixture, and fibers shall be volume proportioned by controlled, calibrated, screw conveyor, or other methods of feed, provided uniform proportions are obtained. The equipment shall be capable of controlling the delivery of material so that inaccuracies do not exceed 1 percent for water; 1-1/2 percent for cementitious materials; 2 percent for sand and coarse aggregate; and 3 percent for admixture and fibers. If these limits cannot be consistently met, then batching shall be by direct weighing.

To ensure accurate, consistent proportioning of aggregate, augers used in dry-mix process shotcrete delivery equipment shall be constructed of abrasion-resistant material to prevent rapid and excessive wear. Auger feed shall be recalibrated as necessary to keep them within the prescribed batching accuracy percent.

The percentage of surface moisture in the sand (ASTM C 566) shall be 3 to 6 percent, by weight, and shall be controlled within this range as may be necessary to maintain uniform feed and to avoid choking the delivery equipment.

Shotcrete batches containing cementitious materials that have been in contact with damp aggregate or other moisture for more than 2 hours shall be wasted at the Contractor's expense.

Batching for the wet-mix process shall be in accordance with the requirements for concrete batching under subparagraph 3.6.g.

When batching with fibers the Contractor shall obtain a good blend of fibers throughout the shotcrete. The Contractor shall furnish appropriate equipment or develop a suitable technique for dispersing the fibers in the mixer free of fiber clumps. A suggested guide is ACI 544.3R.

(4) Mixing. - Mixing for the wet-mix and the dry-mix process shall be as specified in this paragraph. Cementitious materials, sand, coarse aggregate, admixtures, and steel fibers shall be uniformly added and thoroughly mixed by machine before being fed into the delivery equipment.

Mixers used to mix dry ingredients shall discharge the batch without segregation. Mixers shall be tested for uniformity of coarse aggregate content from front to back of the mixer. The maximum permissible difference in percentage of coarse aggregate by weight of sample shall not exceed 6 percent within a batch.

The discharge nozzle for the dry-mix process shall be equipped with a manually operated water injection system of sufficient pressure to provide an even distribution of water into the dry shotcrete mixture at the nozzle. The water valve shall be capable of ready adjustment to vary the quantity of water and shall be convenient to the nozzleman.

(5) Placing. - Placing shotcrete shall be performed only by a nozzleman certified in accordance with subparagraph 3.5.c.(3) during preapplication testing.

An air compressor with ample capacity to provide clean, dry air and maintain a uniform nozzle velocity shall be used for applying shotcrete.

The shotcrete shall be applied by pneumatic pressure from a discharge nozzle held about 2 to 5 feet from the surface and in a stream as nearly normal as possible to the surface being covered. The nozzle shall also be rapidly gyrated while applying the shotcrete.

The shotcrete shall be applied in layers having a thickness that will ensure complete adherence of the shotcrete to the surface. Any shotcrete that shows evidence of sloughing or separation shall be removed and replaced by and at the expense of the Contractor and to the satisfaction of the Contracting Officer.

Care shall be taken to prevent the formation of sand pockets in the shotcrete. Any sand pockets formed shall be removed immediately and replaced with suitable shotcrete at the expense of the Contractor.

Use of rebound as shotcrete aggregate is not permitted, and rebound accumulations shall be removed and disposed of at the expense of the Contractor as approved by the Contracting Officer.

The temperature of shotcrete, as placed, shall be between 50 and 90 °F. Shotcrete shall not be applied to frozen surfaces. The applied shotcrete shall be kept at a temperature of at least 50 °F for a minimum of 3 days immediately following application. When cold weather conditions prevail at the job site and the temperature of aggregates and water is below 50 °F, it may be necessary to heat the aggregate and/or water prior to use in the shotcrete to obtain shotcrete meeting the specified 28-day compressive strength.

The Contractor shall provide and maintain sufficient standby equipment to assure continuous production and application of shotcrete.

If, in the Contracting Officer's opinion, the shotcreting system selected by the Contractor fails to provide satisfactory in-place shotcrete in accordance with these specifications, the Contractor shall change to another system of either of the two processes, provide a redemonstration of the nozzleman's proficiency, or provide a new qualified nozzleman.

If, in the Contracting Officer's opinion, the projection of steel fibers (when used) on the face of the finished shotcrete creates a hazard to personnel, a sacrificial coating of no less than 1/2-inch thickness of shotcrete without the steel fibers shall be applied.

h. Curing. - Shotcrete that is applied where the ambient relative humidity is 85 percent or above will not require measures to control the evaporation of water during curing. However, the Contractor shall substantiate that the relative humidity level in the area of application is above 85 percent by furnishing, installing, and maintaining equipment capable of continuously recording relative humidity.

When the relative humidity is less than 85 percent, the Contractor shall initiate an approved curing method immediately after application of the shotcrete.

Curing shall be accomplished by either:

(1) Raising and maintaining the ambient relative humidity above 85 percent, or

(2) Applying a membrane curing compound as specified in subparagraph 3.6.d.(6).

Water curing. - Shotcrete cured with water shall meet the applicable requirements of subparagraph 3.6.h. except the 14-day requirement may be reduced to 7 days.

## 3.6    CONCRETE REPLACEMENT

a.  General. - Concrete replacement shall be used on areas of damaged or unacceptable concrete greater than 1 square foot having a depth greater than 6 inches or a depth extending 1 inch below or behind the backside of reinforcement.  Concrete replacement shall also be used for holes extending entirely through concrete sections and for large areas of repair greater than 4 inches in depth when the concrete to be repaired is less than 7 days old.  Epoxy bonding agents, latex bonding agents, dry neat cement, cement paste, or cement and sand mortar shall not be used to bond fresh concrete to concrete being repaired by this method.

b.  Submittals. - The Contractor shall submit certification of compliance for materials in accordance with subparagraph d. below.

c.  Quality assurance. - Quality assurance shall be in accordance with paragraph 1.4.

d.  Materials. - All concrete materials shall be obtained from previously tested and approved sources.  Materials will be accepted on certificate of compliance with the following ASTM Standards:

(1)  Portland cement. - Portland cement shall meet the requirements of ASTM C 150 for type I, II, or V cement.  The specific cement type shall be as directed by the Contracting Officer and determined by the environment in which the repair is conducted.

(2)  Pozzolan. - Pozzolan shall meet the requirements of ASTM C 618 for class F pozzolan.

(3)  Admixtures. - The Contractor shall furnish air-entraining and chemical admixtures for use in concrete.

(a)  Air-entraining admixture shall be used in all concrete and shall conform to ASTM C 260.

(b)  Chemical admixtures. - The Contractor may use type A, D, F, or G chemical admixtures.  If used, they shall conform to ASTM C 494.

(4)  Water. - The water used in making and curing concrete shall be free from objectionable quantities of silt, organic matter, salts, and other impurities.

(5) Aggregate. - The term "sand" is used to designate aggregate in which the maximum size particle will pass a 4.75-mm (No. 4) sieve. The term "coarse aggregate" is used to designate all aggregate which can be retained on a 4.75-mm (No. 4) sieve. Sand and coarse aggregate meeting the requirements of ASTM C 33 shall be used in all concrete.

(6) Curing compound. - Wax-base (type I) and water-emulsified resin-base (type II) curing compounds shall conform to the requirements of Reclamation's "Specifications for Concrete Curing Compound" (M-30) dated October 1, 1980.

e.  Safety. - All work shall be performed in accordance with paragraph 1.6.

f.  Concrete preparation. - After damaged or unacceptable concrete has been removed as specified in section 2. the surface on which the replacement concrete will be placed shall be prepared. An acceptable surface shall have the appearance of freshly broken, properly cured concrete. The surface shall be free of any deleterious materials such as free moisture, ice, petroleum products, mud, dust, carbonation, and rust. The perimeters of the repair shall be saw cut to a minimum depth of 1 inch.

The clean surface is not ready to receive repair concrete until it has been brought to a saturated, surface-dry condition. This condition is attained by saturating the surface to a depth that no concrete mixture water may be absorbed from the fresh concrete. Then, just prior to placing concrete against the surface, all free moisture (moisture capable of reflecting light) shall be removed from the prepared surface.

g.  Application. - Replacement concrete shall be composed of cement, coarse aggregate, sand, water, and approved admixtures, all well mixed and brought to the proper consistency. Concrete mixtures shall be proportioned in accordance with Reclamation's "Concrete Manual", Eighth Edition, revised, chapter III. The water-cement ratio of the concrete (exclusive of water absorbed by the aggregates) shall not exceed 0.47 by weight. Slump of the concrete, when placed, shall not exceed 2 inches for concrete in slabs that are horizontal or nearly horizontal and 3 inches for all other concrete. Concrete with less slump should be used when it is practicable to do so. The concrete ingredients shall be thoroughly mixed in a batch mixer. The concrete, as discharged from the mixer, shall be uniform in composition and consistency from batch to batch.

(1) Forms. - Forms shall be used for concrete whenever necessary to confine the concrete and shape it to the required lines. The forms shall be clean and free from encrustations of mortar, grout, or other foreign material. Before concrete is placed, the surfaces of the forms shall be coated with a form oil that will effectively prevent sticking and will not soften or stain the concrete surfaces or cause the surfaces to become chalky or dust producing.

(2) Placing. - Placing of concrete shall be performed only in the presence of an authorized representative of the Contracting Officer. Placement shall not begin until all preparations are complete and the authorized representative of the Contracting Officer has approved the preparations. Concrete shall not be placed in standing or running water unless, as determined by the Contracting Officer, the structure under repair cannot be economically dewatered. If underwater concrete placement is required, special placing procedures shall be required. A suggested guide is ACI 394R.

When appropriate, concrete shall be placed in layers not greater than 20 inches thick. Each layer, regardless of the thickness, shall be adequately consolidated using immersion-type vibrators or form vibrators when approved. Adequate consolidation of concrete is obtained when all undesirable air voids, including the air voids trapped against forms and construction joints, have been removed from the concrete.

(3) Finishing. - The class of finish required shall be a finish closely resembling the finish of the surrounding concrete.

h. Curing and protection. - Concrete repairs shall be cured either by water curing or by use of wax-base (type I) or water-emulsified resin-base (type II) curing compound meeting the requirements of subparagraph 3.6.d.(6). Daily inspection by the Contractor shall be performed to ensure the maintenance of a continuous, water-retaining film over the repaired area. The water-retaining film shall be maintained for 28 days after the concrete has been placed.

Water curing shall commence when the concrete has attained sufficient set to prevent detrimental effects to the concrete surface. The concrete surface shall be kept continuously wet for 14 days.

The Contractor shall protect all concrete against damage until acceptance by the Government. Whenever freezing temperatures are imminent, the Contractor shall maintain the newly placed repair concrete at a temperature of not less than 50 °F for 72 hours. Water-cured concrete shall be protected from freezing for the duration of the curing cycle and an additional 72 hours after the water is removed.

3.7    EPOXY-BONDED EPOXY MORTAR

a. General. - Epoxy-bonded epoxy mortar is defined as freshly mixed epoxy mortar (sand with epoxy binder) that is placed over an epoxy resin bond coat on hardened existing concrete. Epoxy-bonded epoxy mortar repair may be used when the depth of repair is 1-1/2 inches or less. This method may also be used for repair of areas with a depth greater than 1-1/2 inches when those areas are small (less than 1 square foot) and few in number, and where it is impractical to use epoxy-bonded concrete.

Epoxy-bonded epoxy mortar may be used only when:

(1) Moisture in the structure will not collect behind the bond coat and cause damage upon freezing, and

(2) The repair will not be subjected to extremes of temperatures such as those caused by exposure to direct sunlight, extremes of climate, or extremes in water temperature.

All epoxy-bonded epoxy mortar repairs to new construction shall be performed after 7 days from the original placement.

b.  Submittals. -

(1) When directed by the Contracting Officer, the Contractor shall submit samples of the epoxy-resin bonding system.  The samples shall be submitted at least 30 days prior to use in the work to the Bureau of Reclamation, Attn D-8180, Building 56, Denver Federal Center, West Sixth Avenue and Kipling Street, Denver CO 80225.

(2) Certification of epoxy-bonding agent. - The Contractor shall furnish the Contracting Officer the manufacturer's certification of conformance of the epoxy-resin bonding system with these specifications. The certification shall identify the Reclamation solicitation/ specifications number(s) under which the epoxy is to be used and shall include the quantity represented, the batch numbers of the resin and curing agent, and the manufacturer's results of tests performed on the particular combination of resin and curing agent.

c.  Quality assurance. - Quality assurance shall be in accordance with paragraph 1.4.

d.  Materials. -

(1) Epoxy resin. - The same epoxy resin system shall be used for both the bond coat and the epoxy mortar.  The epoxy resin shall meet the requirements of specifications ASTM C 881 for a type I , grade 2, class B or C or a type III, grade 2, class B or C epoxy system.  In addition, it shall be a 100-percent solids system, and no unreactive diluents, wetting agents, or volatile solvents shall be incorporated.

(2) Sand. - The sand for epoxy mortar shall be clean, dry, well-graded sand composed of sound particles passing a 1.18-mm (No. 16) sieve and conform to the following limits:

| Sieve | Individual percent, by weight, retained on sieve |
|---|---|
| 600 $\mu$m (No. 30) | 26 to 36 |
| 300 $\mu$m (No. 50) | 18 to 28 |
| 150 $\mu$m (No. 100) | 11 to 21 |
| pan | [1]25 to 35 |

[1]Range shown is applicable when 60 to 100 percent of pan is retained on the 75 $\mu$m (No. 200) sieve. When 0 to 59 percent of pan is retained on the 75 $\mu$m (No. 200) sieve, the percent pan shall be within the range of 10 to 20 percent, and the individual percentages retained on the 600 $\mu$m, 300 $\mu$m, and 150 $\mu$m (Nos. 30, 50, and 100) sieves shall be increased proportionately.

Sand of this grading is not usually commercially available and may have to be produced by the Contractor. Starting with a concrete sand, the oversized particles shall be removed with a 1.18-mm (No. 16) sieve. Individual sieve sizes of sand can be purchased to mix with the remaining sand to meet the required grading. Most sands require at least the addition of more pan material to meet the required grading.

When directed, minor adjustments in sand grading shall be made to provide a suitable epoxy mortar. Other fillers or commercially available sand gradings prepared specifically for epoxy mortars may be used in epoxy mortar on approval by the Contracting Officer.

The sand shall be maintained in a dry area at no less than 60 °F temperature for 24 hours immediately prior to time of use.

e. Safety. - All work shall be performed in accordance with paragraph 1.6 of these specifications. Certain additional safety precautions shall be employed when using uncured epoxy materials. Skin contact with uncured epoxy shall be avoided. Protective clothing, including rubber or plastic gloves, shall be worn by all persons handling epoxy materials. All exposed skin areas that may come in contact with the material shall be protected with a protective barrier cream formulated for that purpose. Adequate ventilation shall be provided and maintained at all times during use of epoxy and epoxy solvents. Fans used for ventilating shall be explosion proof. If necessary, respirators that filter organic fumes and mists shall be worn. If spray application is used, the operator shall wear a compressed air-fed hood, and no other personnel shall be closer than 100 feet if downwind of the operator when spraying is being performed. All epoxy-contaminated materials such as wipes, empty containers, and waste material shall be continually disposed of in containers which are protected from spillage. Epoxy spillage shall be immediately and thoroughly cleaned up. Appropriate solvents may be used to clean tools and spray guns, but in no case shall the solvents be incorporated in any epoxy resin or in the placing operation. Solvents shall not be used to remove epoxy materials from skin. Only soap, water, and rags shall be used for this purpose.

All tools shall be completely dried after cleaning and before reuse. All materials, tools, and containers contaminated with epoxy resin or epoxy curing agent shall be removed from the site for disposal in accordance with appropriate local or Federal regulations.

f. Concrete preparation. - Surface preparation and needed removal of existing concrete shall be as in section 2, except that the saw cut shall be 1 inch or equal to the depth of the repair, whichever is less. For repairs less than 1 square foot in area, the required vertical edge may be accomplished with a pneumatic tool or hydrodemolition equipment in lieu of saw cutting. For minor cosmetic repairs of surface defects less than 2 inches in diameter, surface preparation may be limited to cleaning with a small wire brush, removing dust, and heating in depth. Epoxy paste meeting the requirements of ASTM C 881, grade 3, may be used without a bond coat for minor cosmetic repairs of surface defects less than 2 inches in diameter in lieu of epoxy-bonded epoxy mortar when all of the conditions for use of epoxy mortar are met. Any overfilling of minor surface defects shall be removed by grinding after hardening is complete. Where repairs are exposed to public view, color matching of the repair material to the existing concrete shall be done.

The surfaces of the existing concrete to which epoxy mortar is to be epoxy bonded shall be prepared and maintained in a clean and dry condition. Unless epoxy mortar application to wet concrete surfaces is approved by the Contracting Officer. The existing concrete shall be preheated in depth. Preheating shall be sufficient to drive internal moisture from the repair surface and prevent its return until the bond coat is in place. Preheating shall not cause damage to or instant setting of the bond coat.

g. Application. -

(1) Forms. - Forms shall be used as necessary to prevent slumping or sagging of finished epoxy-bonded epoxy mortar. Such forms shall be covered with polyethylene film, and form oil shall not be used.

(2) Preparation of epoxy resin for bond coat. - The epoxy resin is a two-component material which requires combination of components and mixing prior to use. Once mixed, the material has a limited pot life and must be used immediately. The bonding system shall be prepared by adding the curing component to the resin component in the proportions recommended by the manufacturer, followed by thorough mixing. Since the working life of the mixture depends on the temperature (longer at lower temperature, much shorter at higher temperature), the quantity to be mixed at one time shall be applied and topped within approximately 30

minutes.  The addition of thinners or diluents to the resin mixture shall not be done.  Both components of class C epoxy shall be stored above 60 °F prior to use.

(3)  Application of epoxy resin bond coat for epoxy-bonded epoxy mortar. - Immediately after the epoxy resin is mixed, it shall be applied to the prepared, dry existing concrete at a coverage of not more than 80 square feet per gallon, depending on surface conditions.  The area of coverage per gallon of agent depends on the roughness of the surface to be covered and may be considerably less than the maximum specified.  The epoxy resin may be applied by any convenient, safe method such as squeegee, brushes, or rollers, which will yield an effective coverage, except that spraying of the materials will be permitted only if an efficient airless spray is used and when the concrete surfaces to receive the agent are 70 °F or warmer, which spray shall be demonstrated as providing an adequate job with minimum overspray prior to approval of its use.

Care shall be exercised to confine the epoxy resin to the area being bonded and to avoid contamination of adjacent surfaces.  However, the epoxy bond coat shall extend slightly beyond the edges of the repair area.

Steel to be embedded in epoxy mortar shall be coated with epoxy resin. The steel shall be prepared in accordance with the requirements of section 2 and by removing all loose rust either with a wire brush or by wet sandblasting.  The exposed steel shall be completely coated with epoxy resin at the time it is being applied to the concrete surfaces of the repair area.

The applied epoxy resin film shall be in a fluid condition at the time the epoxy mortar is placed.  The epoxy resin may be allowed to stiffen to a very tacky condition rather than a fluid condition before epoxy mortar is placed on steep sloping or vertical surfaces, in which case special care shall be taken to thoroughly compact the epoxy mortar against the stiffening bond coat.  In the event the bond coat is curing too quickly to meet the placement requirements, a second bond coat shall be applied over the first while the first bond coat is still tacky.  If any bond coat has cured beyond the tacky state, it shall be completely removed by sandblasting, and proper cleanup, heating, and drying shall be accomplished and a new bond coat applied.

(4)  Placing and finishing. - The epoxy mortar shall be composed of sand and epoxy resin suitably blended to provide a stiff, workable mixture. The epoxy components shall be mixed thoroughly prior to the application of the bond coat and prior to the addition of the sand.  The mixture proportions shall be established, batched, and reported on a weight basis, provided that the dry sand and mixed epoxy may be batched by volume using suitable measuring containers that have been calibrated on a weight basis.  If equivalent volume proportions are being used, care

shall be taken to prevent confusing them with weight proportions. Epoxy
mortar will require approximately 5-1/2 to 6 parts of graded sand to
1 part epoxy, by weight. The Contracting Officer will determine, and
adjust where necessary, the mix proportions for the particular epoxy and
sand being used. The epoxy mortar shall be thoroughly mixed with a
slow-speed mechanical stirrer or other equipment producing equivalent
results. The mortar shall be mixed in small-sized batches so that each
batch will be completely mixed and placed within approximately
30 minutes from the time the two components for the epoxy resin are
combined. The addition of thinners or diluents to the mortar mixture
will not be permitted.

The prepared epoxy mortar shall be tamped, flattened, and smoothed into
place in all areas while the epoxy bond coat is still in a fluid
condition. The mortar shall be worked to grade and given a steel trowel
finish. Special care shall be taken at the edges of the area being
repaired to ensure complete filling and leveling and to prevent the
mortar from being spread over surfaces not having the epoxy bond coat
application. Steel troweling shall be performed in a manner to best
suit the prevailing conditions but, in general, shall be performed by
applying slow, even strokes. Trowels may be heated to facilitate the
finishing. The use of thinner, diluents, water, or other lubricant on
placing or finishing tools will not be permitted, except for final
cleanup of tools. After leveling of the epoxy mortar to the finished
grade where precision surfaces are required, the mortar shall be covered
with plywood panels smoothly lined with polyethylene sheeting and
weighted with sandbags or otherwise braced, or by other means acceptable
to the Contracting Officer, until danger from slumping has passed. When
polyethylene sheeting is used, no attempt shall be made to remove it
from the epoxy mortar repair before final hardening.

Epoxy-bonded epoxy mortar repairs shall be finished to the plane of the
surfaces adjoining the repair areas. The final finished surfaces shall
match the texture of the surfaces adjoining the repair areas.

h.  Curing and protection. - Epoxy-mortar repairs shall be cured
immediately after completion of each repair area at not less than 60 °F
until the mortar is hard. Post curing shall then be initiated at elevated
temperatures by heating, in depth, the epoxy mortar and the concrete
beneath the repair. Post curing shall continue for a minimum of 4 hours at
a surface temperature not less than 90 °F nor more than 110 °F, or for a
minimum of 24 hours at a surface temperature not less than 60 °F nor more
than 110 °F. The heat shall be supplied by using portable propane-fired
heaters, infrared heat lamps, or other approved methods capable of
producing the required temperature and positioned so that the required
surface temperatures are obtained.

In no case shall epoxy-bonded epoxy mortar be subjected to moisture until
after the specified post curing has been completed.

## 3.8   EPOXY-BONDED CONCRETE

a.   General - Epoxy-bonded concrete is defined as freshly mixed portland cement concrete that is placed over a fluid epoxy resin bond coat on hardened existing concrete.  Epoxy-bonded concrete repair may be used when the depth of repair is 1-1/2 inches or greater.

b.   Submittals. -

(1)   When directed by the Contracting Officer, the Contractor shall submit samples of the epoxy-resin bonding system.  The samples shall be submitted at least 30 days prior to use in the work to the Bureau of Reclamation, Attn D-3731, Building 56, Denver Federal Center, West Sixth Avenue and Kipling Street, Denver CO 80225.

(2)   Certification of epoxy-bonding agent. - The Contractor shall furnish the Contracting Officer the manufacturer's certification of conformance of the epoxy-resin bonding system with these specifications. The certification shall identify the Reclamation solicitation/ specifications number(s) under which the epoxy is to be used and shall include the quantity represented, the batch numbers of the resin and curing agent, and the manufacturer's results of tests performed on the particular combination of resin and curing agent.

c.   Quality assurance. - Quality assurance shall be in accordance with paragraph 1.4.

d.   Materials. -

(1)   Concrete materials. - The materials and procedures used to prepare and mix concrete for epoxy-bonded concrete repair shall be as specified in subparagraph 3.6.d. except that slump of concrete, when placed, shall not exceed 1-1/2 inches.

(2)   Epoxy resin. - The epoxy-resin bonding system shall meet the requirements of specification, ASTM C 881 for a type II, grade 2, class B or C epoxy system.  In addition, it shall be a 100-percent solids system and shall not contain unreactive diluents or wetting agents. Volatile solvents shall not be incorporated into the epoxy system.

e.   Safety. - All work shall be performed in accordance with paragraph 1.6 and these specifications.  Certain additional safety precautions shall be employed when using uncured epoxy materials.  Skin contact with uncured epoxy shall be avoided.  Protective clothing, including rubber or plastic gloves, shall be worn by all persons handling epoxy materials.  All exposed skin areas that may come in contact with the material shall be protected with a protective barrier cream formulated for that purpose.  Adequate ventilation shall be provided and maintained at all times during use of epoxy and epoxy solvents.  Fans used for ventilating shall be explosion proof.  If necessary, respirators that filter organic fumes and mists shall be worn.  All epoxy-contaminated materials such as wipes, empty containers, and waste material shall be continually disposed of in containers which are protected from spillage.  Epoxy spillage shall be immediately and thoroughly cleaned up.  Appropriate solvents may be used to clean tools and spray guns, but in no case shall the solvents be incorporated in any epoxy resin or in the placing operation.  Solvents shall not be used to remove epoxy materials from skin.  Only soap, water, and rags shall be used for this purpose.

All tools shall be completely dried after cleaning and before reuse.

All materials, tools, and containers contaminated with epoxy resin or epoxy curing agent shall be removed from the site for disposal in accordance with appropriate local or Federal regulations.

f.  Concrete preparation. - Concrete to be repaired by epoxy-bonded concrete shall be prepared in accordance with the provisions of section 2, except that the perimeters of the repair shall be saw cut to a minimum depth of 1 inch.  Epoxy-bonded concrete shall not be applied to concrete surfaces at a surface temperature less than 60 °F nor greater than 90 °F.

g.  Application of epoxy-bonded concrete. -

(1)  Forms. - Forms shall be used for epoxy-bonded concrete whenever necessary to confine the concrete and shape it to the required lines. The forms shall have sufficient strength to withstand the pressure resulting from placing operations, shall be maintained rigidly in position, and shall be sufficiently tight to prevent loss of mortar from the concrete.

(2)  Preparation of epoxy resin. - The epoxy resin is a two-component material which requires combination of components and mixing prior to use.  Once mixed, the material has a limited pot life and must be used immediately.  The bonding system shall be prepared by adding the curing component to the resin component in the proportions recommended by the manufacturer, followed by thorough mixing.  Since the working life of the mixture depends on the temperature (longer at lower temperature, much shorter at higher temperature), the quantity to be mixed at one time shall be applied and topped within approximately 30 minutes.  The addition of thinners or diluents to the resin mixture will not be permitted.  Both components of class C epoxy shall be at 60 °F prior to use.

(3)  Application of epoxy resin bond coat for surface repairs. - Immediately after the epoxy resin is mixed, it shall be applied to the prepared, dry existing concrete at a coverage of not more than 80 square feet per gallon, depending on surface conditions.  The area of coverage per gallon of agent depends on the roughness of the surface to be covered and may be considerably less than the maximum specified.  The epoxy resin may be applied by any convenient, safe method such as

squeegee, brushes, or rollers, which will yield an effective coverage, except that spraying of the materials will be permitted only if an efficient airless spray is used and when the concrete surfaces to receive the agent are 70 °F or warmer, which spray shall be demonstrated as providing an adequate job with minimum overspray prior to approval of its use. If spray application is used, the operator shall wear a compressed air-fed hood, and no other personnel shall be closer than 100 feet if downwind of the operator when spraying is being performed.

Care shall be exercised to confine the epoxy resin to the area being bonded and to avoid contamination of adjacent surfaces. However, the epoxy bond coat shall extend slightly beyond the edges of the repair area.

Steel to be embedded in epoxy-bonded concrete shall be coated with epoxy resin. The steel shall be prepared in accordance with the requirements of section 2 and in the same manner required for preparation of the concrete being repaired. The exposed steel shall be completely coated with epoxy resin at the time it is being applied to the concrete surfaces of the repair area.

The applied epoxy resin film shall be in a fluid condition at the time the concrete is placed, provided that the epoxy resin may be allowed to stiffen to a very tacky condition rather than a fluid condition before concrete is placed on steep sloping or vertical surfaces, in which case special care shall be taken to thoroughly compact the concrete against the stiffening bond coat. In the event that an applied film cures beyond the fluid condition, or a very tacky condition where permitted, before the concrete is placed, a second bond coat shall be applied while the first bond coat is still tacky. If any bond coat has cured beyond the tacky state, it shall be completely removed by sandblasting, and proper cleanup, heating, and drying shall be accomplished, and a new bond coat applied.

(4) Placing and finishing. - Use of epoxy-bonded concrete in repairs requiring forming, such as on steeply sloped or vertical surfaces, will be permitted only when the forming required is such that the bond coat can be applied and the concrete properly placed within the time period necessary to ensure that the applied bond coat will still be fluid, or tacky where permitted.

Immediately after application of the epoxy bond coat, while the epoxy is still fluid, concrete shall be spread evenly to a level slightly above grade and compacted thoroughly by vibrating, tamping, or both. Vibrators shall not be permitted to penetrate through the fresh concrete to the level of the fluid epoxy bond coat. Such vibration can emulsify the epoxy and reduce the bond. Tampers shall be sufficiently heavy for thorough compaction. After being compacted and screeded, the concrete shall be given a wood float or steel trowel finish, as directed. Troweling, if required, shall result in a smooth, dense finish that is free from defects and blemishes. As the concrete continues to harden, the surface shall be given successive trowelings. The final troweling shall be performed after the surface has hardened to such an extent that no cement paste will adhere to the edge of the trowel. The number of trowelings and time at which trowelings are performed shall be subject to approval of the Contracting Officer.

The surfaces of epoxy-bonded concrete repairs shall be finished to the plane of the surfaces adjoining the repair areas. The final finished surfaces shall match the texture of the surfaces adjoining the repair areas.

h.  Curing and protection. - The Contractor shall cure and protect all repairs from damage until acceptance by the Government. Concrete shall be protected against freezing for not less than 6 days from time of placement.

As soon as the epoxy-bonded concrete has hardened sufficiently to prevent damage, the surface shall be moistened by spraying lightly with water and then covering with sheet polyethylene, or by applying an approved curing compound, provided that curing compound shall be used for curing concrete whenever there is any possibility that freezing temperatures will prevail during the curing period. Sheet polyethylene, if used, shall be an airtight, nonstaining, waterproof covering which will effectively prevent loss of moisture from the concrete by evaporation. Edges of the polyethylene shall be lapped and sealed. The waterproof covering shall be left in place for not less than 14 days.

If a waterproof covering is used and the concrete is to be subjected to any use that might rupture or otherwise damage the covering during the curing period, the covering shall be protected by a suitable layer of clean wet sand or other cushioning material that will not stain concrete, as approved by the Contracting Officer. Application of curing compound, if used, shall be in accordance with Reclamation's "Specifications for Concrete Curing Compound, M-30." After the curing has been accomplished, the covering, except curing compound if used, and all foreign material shall be removed and disposed of as directed.

3.9   POLYMER CONCRETE SYSTEMS

a. General. - Polymer concrete repair systems for concrete may be either of two types: a methacrylate monomer system, or a vinyl ester resin system. These materials may be used for patches, overlays, grout pads, and embedment of sills, gates, and similar structures in concrete members. Prior approval of the Contracting Officer is required for applications where the repair is exposed to the direct rays of the sun or subjected to rapid temperature changes in excess of 10 °F per hour for more than a 3-hour period. The materials may be supplied as a prepackaged system or as components for a system designed for use as polymer concrete.

The polymer concrete system shall consist of a 100-percent reactive monomer or resin system (no nonreactive diluents or solvents are permitted); an initiator for polymerization of the resin or monomer system, a promoter to activate the initiator; aggregate; and a primer system to be applied to the surface of the concrete to be repaired. The system shall be supplied with or without pigments to approximate the color of concrete, as directed by the Contracting Officer.

b. Submittals. - Before starting work, the Contractor shall submit to the Contracting Officer for approval the following documents:

(1) A safety plan.

(2) A statement of technical qualifications, training, and past experience in handling and applying polymer concrete materials.

(3) A manufacturer's affidavit that states the chemical constituents and proportions of the material, a Materials Data Safety Sheet for each component, the use for which the material is designed, instructions on storage and use of the materials, and typical mechanical and physical properties of the final product.

c. Quality assurance. - Quality assurance shall be in accordance with paragraph 1.4.

d. Materials. -

(1) Monomer and resin systems. -

(a) Methacrylate monomer systems. - The monomer system may be based on either methyl methacrylate monomer or on high molecular weight methacrylate monomer systems. The monomer system shall consist of 100-percent reactive components, and no nonreactive diluents or solvents shall be permitted. The initiator shall be an organic peroxide. The promoter shall be either a cobalt salt or an organic amine compound. Materials shall be used in the proportions recommended by the manufacturer for the temperature conditions at the job site to meet the pot life and curing time requirements of these specifications.

(b) Vinyl ester resin systems. - The vinyl ester resin shall be an elastomer-modified dimethacrylate diglycidyl ether of bisphenol A and shall be Dow Chemical Company Derakane 8084; or equal having the following salient characteristics:

Liquid Resin Properties
    Styrene content                     45 percent by weight
    Viscosity, Brookfield
        (cps), 77 °F                    1,200
    Flash point, cc, (°F)               82

Clear Casting Properties

| | |
|---|---|
| Tensile strength, ($lb/in^2$) | 10,000 |
| Tensile modulus ($lb/in^2$) | 430,000 |
| Elongation, percent | 10 |
| Flexural strength, ($lb/in^2$) | 16,000 |
| Flexural modulus, ($lb/in^2$) | 420,000 |
| Heat distortion temperature (°F) | 170 |

The initiator shall be an organic peroxide and the promoter shall be cobalt napthenate or cobalt octoate. Materials shall be used in the proportions recommended by the manufacturer for the temperature conditions at the job site to meet the pot life and curing time requirements of these specifications.

(2) Aggregates. -

(a) Prepackaged systems. - Prepackaged systems normally contain a preblended fine aggregate. The use of additional fine aggregate shall not be permitted in these systems unless specifically authorized by the manufacturer. Prepackaged systems that do not contain a preblended fine aggregate shall use a fine aggregate meeting the requirements of subparagraph 3.9.d.(2)(b). Prepackaged polymer concrete systems may be extended by the addition of coarse aggregate. The maximum size of the coarse aggregate shall not exceed the lesser of either one-third of the depth of the repair or 1-1/2 inches. The coarse aggregate shall be composed of hard, dense, clean durable, well-graded rock particles.

(b) Separate component polymer concrete systems. - The aggregate shall be composed of hard, dense, clean, durable, well-graded rock particles. Not more than 2 percent by weight of the fine aggregate shall pass the 75 $\mu$m (No. 200) sieve, and the maximum aggregate size shall not be greater than one-third the depth of the repair. The grading shall be based on the following guide:

| Nominal size fraction | Aggregate grading percent retained by weight | | | | | |
|---|---|---|---|---|---|---|
| 9.5 - 19.0 mm (3/8 - 3/4 inch) | -- | -- | -- | -- | -- | 29 |
| 4.75 - 9.5 mm (No. 4 - 3/8 inch) | -- | -- | -- | -- | 29 | 20 |
| 2.36 - 4.75 mm (No. 8 - No. 4) | -- | -- | -- | 29 | 21 | 16 |
| 1.18 - 2.36 mm (No. 16 - No. 8) | -- | -- | 31 | 21 | 16 | 10 |
| 600 $\mu$m - 1.18 mm | -- | 31 | 22 | 15 | 10 | 7 |

| Nominal size fraction | Aggregate grading percent retained by weight | | | | | |
|---|---|---|---|---|---|---|
| (No. 30 - No. 16) | | | | | | |
| 300 - 600 μm (No. 50 - No. 30) | 50 | 23 | 16 | 10 | 7 | 5 |
| 150 - 300 μm (No. 100 - No. 50) | 30 | 16 | 10 | 8 | 4 | |
| *Pan | 20 | 30 | 21 | 17 | 13 | 9 |

*NOTE:  Pan material shall consist of:
1.  Minus 75 μm (No. 200) sieve size ground silica, or
2.  Minus 75 μm (No. 200) sieve size crushed hard, dense, durable, clean rock washed free of clay and organic impurities, or
3.  A mixture of 1 or 2 above with 50% fly ash.

(3)  Coupling agent. - The coupling agent incorporated into the monomer or resin system shall be an organosilane compound, Union Carbide A-174, or equal.

(4)  Primer. - The polymer concrete shall be applied to a prepared and primed concrete surface.  The primer shall consist of either a methyl methacrylate monomer system, a high molecular weight methacrylate monomer system, or  an elastomer-modified vinyl ester resin system (Dow Chemical Company Derakane 8084, or equal).  The same monomer or resin system used in the polymer concrete is acceptable as a primer.  The polymer concrete shall be applied to a primed surface while the primer is still tacky and has not completely cured to a hard and dry finish, with the exception of methyl methacrylate polymer concrete systems applied to a methyl methacrylate primer system which does not contain cross-linking comonomers which may be applied to a cured and dry primed surface.  The primer system shall be formulated according to the manufacturer's instructions to give adequate pot life for the job site temperature conditions for proper application of the polymer concrete.

(5)  Polymer concrete properties. - The polymer concrete shall have the following properties:

(a)  Curing time of 1 to 3 hours at ambient temperature (substrate and air) of 32 to 100 °F.

(b)  Pot life of 15 to 30 minutes at ambient temperatures of 32 to 100 °F.

(c)  Compressive strength of at least 7,000 lb/in$^2$.  (ASTM C 39).

(d)  Splitting tensile strength of at least 1,000 lb/in$^2$.  (ASTM C 496).

(e)  Linear shrinkage, less than 0.05%.

(f)  Bond strength to concrete - at least equal to the splitting tensile strength of the base concrete.

e.  Safety. - All work shall be performed in accordance with requirements of paragraph 1.6 and these specifications.  The Contractor shall also:

(1)  Hold a safety meeting at the job site conducted by an industrial hygienist or by a technically qualified professional staff member to acquaint and instruct all workers and supervisors on the job in the proper care and handling of the polymer concrete materials as specified by the manufacturer of the polymer concrete, safety precautions to be observed, personal protective gear, and protection of the environment.

(2)  Require all workers, supervisors, inspectors, visitors, and other people at the job site to wear personal protective gear as directed by the Contracting Officer.

(3)  Ensure that the polymer concrete materials are stored, handled, and applied in the manner specified by the manufacturer.  In addition:

(a)  Storage. - polymer concrete materials shall be stored in the shipping containers in a well-ventilated area and out of the direct rays of the sun.  The storage temperature shall not exceed 80 °F (27 °C).  Materials shall not be stored longer than 3 months.

(b)  Mixing and application. - No smoking, flame, or other ignition sources shall be permitted during mixing and application.  Type B or type ABC fire extinguishers shall be provided at the mixing and application sites.  Electrical equipment in contact with the polymer concrete should be grounded for safe discharge of static electricity.

(c)  Handling. - Workers shall be provided and required to wear rubber boots, disposable protective clothing, splash-type safety goggles, rubber gloves, and organic vapor respirators as directed by the Contracting Officer.  A heated eye wash capable of sustaining a 15-minute stream of clean, room temperature water shall be provided at the mixer and at the application site.  Materials coming into direct contact with the skin shall be immediately removed using soap and water.

(4)  Prevent the contamination of soil or water at the job site by liquid components.

(5) Dispose of liquid components and excess materials at the job site by combining the materials in the same manner and procedure used for mixing polymer concrete, placing the mixed materials in an open container, and allowing the material to harden. Hardened polymer concrete is nonpolluting and may be disposed of as a solid nonhazardous waste.

f. Concrete preparation. - In addition to the requirements of section 2, the concrete surface shall be dry and primed with an approved primer. The primer and the polymer concrete shall not be applied until the surface preparation meets the approval of the Contracting Officer. The dryness of the surface shall be determined by taping a sheet of transparent polyethylene sheeting to the surface and exposing it to the full rays of the sun for at least 4 hours and observing the interior surface of the polyethylene sheeting for the occurrence of condensed moisture. As an alternative method, a calibrated moisture meter which meets the approval of the Contracting Officer may be used to determine the dryness of the surface.

g. Preparation and application requirements. -

(1) Initiator and promoter. - The initiator and promoter shall be prebatched and packaged separately from each other in a manner so that the two components cannot be combined until the time of the concrete mixing operation. The promoter shall be combined with the monomer or resin system prior to the addition of the initiator. UNDER NO CIRCUMSTANCES SHALL THE INITIATOR BE ADDED TO OR DIRECTLY CONTACT THE PROMOTER. THE DIRECT COMBINATION OF PROMOTER AND INITIATOR WILL RESULT IN AN EXTREMELY VIOLENT AND EXPLOSIVE REACTION.

(2) Sequence of addition and combination of materials at the mixer. - The polymer concrete materials shall be combined in the sequence and manner specified by the manufacturer.

(3) Mixing of polymer concrete. - The polymer concrete shall be mixed in a paddle-type power mixer, a rotating drum-type power mixer, or other type of power equipment approved by the Contracting Officer. Polymer concrete materials shall be mixed according to the recommendations of the manufacturer of the polymer concrete materials for at least 3 minutes.

(4) Removal of polymer concrete from the mixer. - The polymer concrete shall be removed from the mixer immediately after mixing.

(5) Appropriate forms shall be used for polymer concrete whenever necessary to confine the polymer concrete and shape it to required lines. The surfaces of the forms shall be coated with a release agent that will effectively prevent sticking without damage to the polymer concrete surfaces.

3.10    THIN POLYMER CONCRETE OVERLAY

   a.   General. - Thin polymer concrete overlay shall consist of one coat of
   primer and one or more coats of sealant as directed by the Contracting
   Officer.

   The coat of primer shall consist of vinyl ester resin, initiator, and
   promoter.  Each coat of sealant shall consist of the same materials as in
   the primer, but with the addition of silica filler, titanium dioxide
   pigment, and carbon black pigment.

   b.   Submittals. - The Contractor shall, before starting work, provide a
   manufacturer's affidavit which indicates the chemical constituents of the
   material by proportion.  The chemical constituents shall correspond to the
   requirement of subparagraph 3.10.d.

   c.   Quality assurance. - Quality assurance shall be in accordance with
   paragraph 1.4 and these specifications.

   The vinyl ester resin shall be stored in the shipping containers and kept
   in the shade, well ventilated, and out of direct sunlight.  The recommended
   storage temperature is 50 to 75 °F.  Resin shall not reach 80 °F or higher,
   as it will start to gel.  The resin has a storage life of about 3 months.
   During storage, the resin shall not come in contact with copper, brass,
   zinc, or rust, as discoloration, polymerization, or interference with
   normal cure conditions can occur.  Resin shall also not contact rubber, as
   resin is a solvent for rubber.

   Styrene monomer, which is a component of vinyl ester resin, is flammable
   and forms explosive mixtures in the air; however, it is not sufficiently
   flammable to be listed as a "flammable liquid" under Interstate Commerce
   Commission definitions (flash point at or below 80 °F).  The explosive
   limits are 1.1 to 6.1 percent volume in air.

   The Contracting Officer will not allow the use of vinyl ester resin if it
   has already polymerized, gelled, discolored, or will not cure under normal
   conditions.  Substandard materials shall be replaced at the expense of the
   Contractor.

   d.   Materials. -

      (1)  Vinyl ester. - The vinyl ester resin material shall be Dow Derakane
      8084 elastomer-modified vinyl ester resin, manufactured by Dow Chemical
      Co., 2800 Mitchell Drive, Walnut Creek CA 94596; or equal.  It shall
      meet the requirements of 3.9.d.(1)(b).

      (2)  Initiator. - The initiator shall be cuemene hydroperoxide - 78
      percent as manufactured by:  Lucidol Division, Penwalt Corp., 1740
      Military Road, Buffalo NY 14240; Reichold Chemicals, Inc., 107 South

Motor Avenue, Azusa CA 91706; Witco Chemicals, U.S. Peroxygen Division, 850 Morton Avenue, Richmond CA 94804; or equal.

(3)  Promoter. - The promoter shall be cobalt napthenate; 6 percent.

(4)  Filler. - The filler shall be ground silica, minus 45 $\mu$m (No. 355) sieve size as manufactured by: Ottawa Silica Co., Ottawa, Illinois; VWR Scientific (Silco Seal 395 Ground Silica), PO Box 3200, San Francisco CA 94119; or equal.

(5)  Pigment for opaqueness. - Two pigments shall be used and they shall be:

(a) Titanium dioxide powder as manufactured by: VWR Scientific, PO Box 3200, San Francisco CA 94119; MCB Manufacturing Chemist, Inc., 470 Valley Drive, Brisband CA 94005; or equal.

(b)  Carbon lamp black or bone black powder. - (Do not use activated carbon)

e.  Safety. -  All work shall be performed in accordance with paragraph 1.6 and these specifications.  No smoking, flame, or other ignition source shall be present during mixing and the application procedures.  Fire extinguishers (types B or ABC) shall be provided.  Equipment contacting resins shall be grounded for safe discharge of static electricity.  Tools used shall be the non-sparking, special alloy type.

Workers shall be provided with rubber boots, rubber gloves, protective clothing, safety goggles or face shields, and organic vapor respirators. It is important to avoid inhalation of vapors and direct eye or skin contact.  Eyewash facility shall be available which will provide a clean, room temperature flushing stream for a minimum of 15 minutes.  Contaminated clothing shall be discarded immediately.  Proper tools and facilities to quickly remove spills shall be at the worksite.

Personnel shall follow manufacturer's recommendations for safe handling of vinyl ester resin and all additives.

CUEMENE HYDROPEROXIDE INITIATOR SHALL NEVER BE ADDED DIRECTLY TO COBALT NAPTHENATE PROMOTER OR AN EXTREMELY VIOLENT AND EXPLOSIVE REACTION WILL OCCUR.

f.  Concrete preparation. - Concrete surfaces designated to receive the thin polymer concrete overlay shall be prepared in accordance with section 2.  The minimum preparation shall consist of wet sandblasting or water blasting the concrete to a clean, sound surface condition followed by drying.

g.  Application of the thin polymer concrete overlay. -

(1) Coverage. - An average coverage rate of 50 to 75 square feet per gallon of applied material per coat over the entire surface is anticipated for this work. The surface texture of the concrete may affect this coverage rate.

(2) Mixing proportion. -

    (a) Primer: 5.0 gallons vinyl ester resin
                  0.60 pound initiator
                  0.25 pound promoter

    (b) Sealant: 5.0 gallons vinyl ester resin
                  1.35 pounds initiator
                  0.27 pounds promoter
                  40.0 pounds filler
                  4.0 pounds titanium dioxide pigment
                  0.02 pound carbon lamp black pigment

(3) Variations. - The proportions of cobalt napthenate promoter and cuemene hydroperoxide initiator were selected to give a 2- to 4-hour pot life at a temperature of about 70 °F. The rate of the polymerization reaction also depends on temperature - in colder weather the reaction will be slower, and in hot weather or in sunshine, the reaction will proceed faster. The temperature effects can be compensated for, to a certain extent, by increasing the total amount of initiator plus promoter in cold weather and decreasing the amount in hot weather.

Great care shall be exercised in proportioning the titanium dioxide and the carbon black to ensure exact color between batches of sealant layer.

The Contracting Officer may require variations from the proportions and quantities specified above and trial mixtures.

At no additional cost to the Government the Contracting Officer may require two trial mixtures utilizing a total of 5 gallons of vinyl ester resin with a proportional amount of other materials as indicated in subparagraph 3.10.g.(2). The Contractor shall supply, mix, and apply the trial polymer overlay where directed and in a manner as indicated by these specifications. The Contracting Officer shall be given 12 hours' notice before a trial mix is applied.

(4) Mixing. - The filler, pigments, and cobalt napthenate promoter shall be mixed with vinyl ester resin in a paddle-type or rotating drum power mixer in advance of the application. Mixtures shall be prepared in maximum volumes of 5 gallons of resin per mixed batch.

The Contractor shall mix the required constituents, with the exception of the cuemene hydroperoxide, to the satisfaction of the Contracting Officer. The sealant material will then be required to set for 45 to 90 minutes for the filler and pigment to become thoroughly wetted by the vinyl ester resin. Just prior to use, the cuemene hydroperoxide initiator shall then be added and the sealant mixture mixed again to the Contracting Officer's satisfaction. CUEMENE HYDROPEROXIDE INITIATOR AND COBALT NAPTHENATE PROMOTOR SHALL NEVER BE ADDED TOGETHER DIRECTLY, OR AN INSTANTANEOUS AND VIOLENTLY EXPLOSIVE REACTION COULD OCCUR. For the primer, the setting period may be omitted, but the initiator must still be added after the first mixing of the promoter, then mixed again.

(5) Application. - The primer shall uniformly and completely cover the surface by being spread and scrubbed into the concrete surface with a paint roller, brush, or push broom. Discontinuities and puddles shall be eliminated by vigorous scrubbing action. The application rate of primer is expected to be 1 gallon per 50 to 75 square feet of surface area, but may vary due to texture of the prepared surface.

The pigmented sealant layers of the overlay material shall be applied not less than 4 and no greater than 24 hours after the application of the primer or succeeding sealant layer. The primer layer may not have completely cured during this time period. The sealant shall be applied to dry, primed surfaces, and provisions shall be taken to prevent wetting of the surfaces from rain or leakage and seepage of water. The sealant layers shall also be applied with a paint roller, brush, or push broom. The application rate of the sealant is expected to be 1 gallon per 50 to 75 square feet of surface area.

The primer and sealant layers shall be spread evenly and uniformly over the surface.

3.11    RESIN INJECTION

a. General. - When hardened concrete is cracked in depth or when hollow plane delaminations or open joints exist in hardened concrete and when structural integrity or watertightness must be restored for the structure to be serviceable, resin injection shall be used for repair, as directed.

However, since not all cracked, delaminated, or jointed concrete can be restored to serviceable condition by resin injection, resin injection repairs shall be made only as directed by the Contracting Officer.

Two basic types of injection resin are used to repair concrete:

(1) Epoxy resins are used to rebond cracked concrete and to restore structural soundness. Epoxy resins may also be used to eliminate water leakage from concrete cracks or joints, provided that cracks to be injected with epoxy resin are stationary. Cracks that are actively leaking water and that cannot be protected from uncontrolled water inflow shall not be injected with epoxy resin. Cracks to be injected with epoxy resin shall be between 0.005 inch and 0.25 inch in width.

(2)  Hydrophilic polyurethane resin is used to eliminate or reduce water leakage from concrete cracks and joints and can be used to inject cracks subject to some degree of movement.  Hydrophilic polyurethane resin shall not be used to inject concrete cracks or joints when restoration of structural bond is desired.  Cracks to be injected with polyurethane resin shall be 0.005 inch in width or greater.

Other types of injection resin are available for nonstandard or specialized repair applications.  Use of these materials shall require specific approval of the Contracting Officer.

b.  Submittals. - The Contractor shall provide the Contracting Officer with evidence that the Contractor is qualified to perform resin injection repairs.  The data shall show that the Contractor has a minimum of 3 years of experience in performing resin injection work similar to that detailed in the drawings and specifications.

The Contractor shall submit a list of 5 projects in which resin injection was successfully completed.  The list shall contain the following information for each project:

(1)  Project name and location.

(2)  Owner of project.

(3)  Brief description of work.

(4)  Date of completion of resin injection work.

If the repair work is performed by the Contractor's personnel under the supervision of the manufacturer's representative, the data shall show that the resin manufacturer has a minimum of 5 years' experience providing resin materials similar to those specified.

Manufacturer's brochures, technical data sheets, Material Safety Data Sheets, and any other information describing the polyurethane resin, the proper formulation to achieve the required tensile strength, bond strength, and elongation of the cured resin mixture, and recommended injection procedures shall be submitted to the Contracting Officer for approval at least 30 days prior to commencement of crack repairing operations.

The Contractor shall furnish the Contracting Officer a manufacturer's certification of conformance of the epoxy or polyurethane resin system with these specifications.  The certification shall identify the Reclamation solicitation/specifications number(s) under which the resin is to be used and shall include the quantity represented, the batch numbers of the resin, and the manufacturer's results of tests performed on the resin system.

The Contractor shall submit a detailed proposal for epoxy injection repair to the Contracting Officer for approval.  The approval will be based on the degree of conformance of the proposal with procedures contained in Reclamation's Concrete Manual, chapter 7, Eighth Edition, revised reprint, and the report of ACI Committee 503, section 7.2.5, "Use of Epoxy Compounds with Concrete."

The Contractor shall submit a detailed proposal for polyurethane resin injection repair to the Contracting Officer for approval. The approval will be based on the degree of conformance to the basic steps of polyurethane resin injection and on the Contracting Officer's judgment of the technical feasibility of the Contractor's proposal.

c.  Quality assurance. - Quality assurance shall be in accordance with paragraph 1.4 and these specifications. The repair work may be performed by the Contractor or by the Contractor's personnel under the supervision of the resin manufacturer's representative. If the Contractor performs the repair work, the Contractor shall provide a full-time, onsite supervisor throughout the duration of the resin injection work.

If the repair work is performed by the Contractor's personnel under the supervision of the resin manufacturer's representative the Contractor shall have the resin manufacturer provide a representative who will train the Contractor's personnel on the proper techniques of injecting resin with an injection system approved by the manufacturer. Also, the Contractor shall provide a full-time, onsite, manufacturer-certified injection supervisor throughout the duration of the resin injection work.

d.  Materials. -

   (1)  Epoxy resin. - Epoxy resin for injection shall meet the requirements of specification ASTM C-881 for a type I, grade 1 epoxy system. The class of the system shall be appropriate for the temperature of the application.

   (2)  Polyurethane resin. - The polyurethane resin system for injection into cracked concrete shall be a two-part system composed of 100 percent hydrophilic polyurethane resin and water. The polyurethane resin, when mixed with water, shall be capable of forming either a flexible closed-cell foam or a cured gel dependent upon the water-to-resin mixing ratio. The amount of water mixed with the polyurethane resin shall be such that the cured material meets the following physical properties:

      (a)  Minimum tensile strength -- 20 pounds per square inch.

      (b)  Bond to concrete (wet) -- greater than 20 pounds per square inch.

      (c)  Minimum elongation -- 400 percent.

   The injection of pure polyurethane resin, not mixed with water, shall not be allowed.

e.  Safety. - All work shall be performed in accordance with paragraph 1.6 and these specifications.

(1)  Epoxy resins. - Certain additional safety precautions shall be employed when using uncured epoxy materials.  Skin contact with uncured epoxy shall be avoided.  Protective clothing, including rubber or plastic gloves, shall be worn by all persons handling epoxy materials.  All exposed skin areas that may come in contact with the material shall be protected with a protective barrier cream formulated for that purpose.  Adequate ventilation shall be provided and maintained at all times during use of epoxy and epoxy solvents.  Fans used for ventilating shall be explosion proof.  If necessary, respirators that filter organic fumes and mists shall be worn.  All epoxy-contaminated materials such as wipes, empty containers, and waste material shall be continually disposed of in containers which are protected from spillage.  Epoxy spillage shall be immediately and thoroughly cleaned up.  Appropriate solvents may be used to clean tools and spray guns, but in no case shall the solvents be incorporated in any epoxy resin or in the placing operation.  Solvents shall not be used to remove epoxy materials from skin.  Only soap, water, and rags shall be used for this purpose.

All tools shall be completely dried after cleaning and before reuse.

(2)  Polyurethane resins. - Polyurethane injection resin systems contain either toluene diisocyanate or methylene dephenyl diisocyanate.  Both isocyanates can create risks if safe handling procedures are not followed.  The principal hazards arise from isocyanate vapor, which will irritate the membranes of the nose, throat, lungs, and eyes.  Adequate ventilation is required to prevent vapor concentrations from approaching the Threshold Limit Value (TLV).  Protective clothing, including rubber or plastic gloves and protective glasses, shall be worn by all persons handling polyurethane resins.  If necessary, respirators that filter isocyanate vapors and mists shall also be worn.  Monomeric urethane resins react with water to produce polyurethane and carbon dioxide gas.  If this reaction occurs inside a closed container, excessive pressures can develop that may rupture the container.  Care must be taken to prevent contamination of monomeric urethane resin with water.

Polyurethane resin spillage shall be immediately and thoroughly cleaned up.  Spilled polyurethane resin can be absorbed in sand and removed for burial.

(3)  Cleanup and disposal of injection resin. - All materials, tools, and containers contaminated with injection resin, surface sealers, or other contaminants shall be removed from the site for disposal in accordance with appropriate local or Federal regulations.

f.  Concrete preparation. - The concrete surface to be repaired by resin injection shall be thoroughly cleaned of all deteriorated concrete, efflorescence, and all other loose material.  The area to be injected shall then be thoroughly inspected and an injection port drilling and pumping pattern established.

Upon completion of resin injection, all excess material shall be removed from the exterior surfaces of the concrete. The final finished surfaces shall match the texture of the surfaces adjoining the repair areas.

g.  Application of resin injection repairs to concrete. -

(1)  Epoxy resin injection repair. - The process used for epoxy injection shall fill the entire crack or hollow plane delamination with liquid epoxy resin system and shall contain the resin system in the crack until it has hardened. The Contractor shall be responsible for drilling and removing three, a minimum of 2-inch-diameter cores from the injected concrete at locations determined by the Contracting Officer to determine the completeness of the injection repair. Injection shall be considered complete if more than 90 percent of the void is filled with hardened epoxy. If injection is not complete, reinjection and additional cores may be required at the direction of the Contracting Officer at no additional cost to the Government.

Epoxy injection repair methods shall be in accordance with the approved detailed proposal for epoxy injection repair and shall be adjusted to fit the repair situations encountered.

(2)  Hydrophilic polyurethane resin injection repairs. - The process used for polyurethane injection of cracks or joints to reduce water leakage shall consist of the following basic steps:

(a)  Intercept the water flow paths with valved drains installed into the concrete to control the leakage.

(b)  Install injection ports by drilling holes designed to intersect the cracks at depth below the concrete surface. The maximum spacing of injection ports shall not exceed 60 inches, and closer spacing of ports may be required.

(c)  All injection holes shall be flushed with clean water to remove drilling dust and loose debris and to clean the intersected crack line. Each drill hole shall be water tested at the resin injection pressure to determine if the crack intersection is open. Polyurethane resin shall not be pumped into a drill hole that refuses to take water at the resin injection pressure.

(d)  Inject polyurethane resin system into cracks or joints at the minimum pressure required to obtain the desired travel, filling, and sealing. The mix water to resin ratio shall be 1:1 unless otherwise approved by the Contracting Officer. The Contractor should anticipate the necessity to provide a surface seal for the crack or joint to contain the injection resin. It may also be necessary to inject the crack or joint in an intermittent manner to achieve filling and sealing. Injection shall be by the method of split spacing unless otherwise approved by the Contracting Officer. Primary holes shall be drilled and injected on centers not exceeding 10 feet. Secondary holes, half way between the primary holes, will then be drilled and injected. If resin take occurs in the secondary holes, a series of tertiary holes, half way between the secondary and primary holes, shall then be drilled and injected. All holes shall be injected to absolute refusal.

(e) Remove drains, injection ports, and excess polyurethane upon completion of resin cure.

This process shall entirely stop the water leakage to a dust dry condition or as directed by the Contracting Officer.

The pump used to inject the polyurethane resin system shall be a two-component positive-displacement-type pump with static mixing head and pressure regulation necessary to control injection pressures while pumping low volumes. The equipment will be subject to approval by the Contracting Officer. The use of single component pumps and/or the injection of pure water followed by injection of pure resin will not be approved.

Polyurethane resin injection methods shall be in accordance with the approved, detailed proposal for injection repair and shall be adjusted to fit the repair situations encountered.

3.12    HIGH MOLUCULAR WEIGHT METHACRYLIC SEALING COMPOUND

a.    General. - A concrete sealing compound is defined as a liquid that is applied to the surface of hardened concrete to prevent or decrease the penetration of liquid or gaseous media, such as water, aggressive solutions, and carbon dioxide, during the service exposure, preferably after initial drying to facilitate its absorption into voids and cracks.

The sealing of concrete surfaces with a high molecular weight methacrylic monomer-catalyst system and sand shall be in accordance with these specifications. Other types of concrete sealing compounds may be used by the Contractor only when approved by the Contracting Officer.

The high molecular weight methacrylic sealing compound shall not be used to seal concrete subject to frequent or permanent immersion in water; nor shall it be used on concrete surfaces exposed to high abrasion forces.

b.    Submittals. - The Contractor shall provide the Contracting Officer a table showing preparation of initiator and promoter to be added to the monomer to achieve the cure time requirements based on concrete surface temperature. The temperature of the surface to be treated shall range from

45 to 100 °F.  If it is desired to work outside these temperature ranges, approval of the Contracting Officer is required, and the monomer manufacturer should be consulted for technical advice.

A MSDS shall be furnished to the Contracting Officer prior to shipment of material with information pertaining to the safe practices for storage, handling and disposal of the materials and their explosive and flammable characteristics, health hazards, and the manufacturer's recommended fire fighting techniques.  The MSDS shall be posted at all storage areas and at the job site.

The Contractor shall furnish, for approval, a detailed written description of the methods the Contractor plans to use for clean up of all spills and residue in open containers.  See subparagraph 3.12.g.(4).

c.  Quality assurance. - Quality assurance shall be in accordance with paragraph 1.4.

d.  Materials. -

(1)  High molecular weight methacrylic monomer. - The monomer shall be a high molecular weight or substituted methacrylate that conforms to the following properties:

(a)  Vapor pressure    Less than 1.0 mm HG @ 77 °F  (ASTM D 323)

(b)  Flash point       Greater than 200 °F  (Pensky-Martens CC)

(c)  Density           Greater than 8.4 lb/gal at 77 °F  (ASTM D 2849)

(d)  Viscosity         12 ± 4 cps (Brookfield No. 1 Spindle 60 rpm, at 73 °F)

(e)  Index of refraction    1.470 ± 0.002

(f)  Boiling point @ 1 mm Hg, degrees F 158

(g)  Shrinkage on cure -less than- 11%

(h)  Glass transition temperature (DSC), degrees F. 135 °F (ASTM D 3418)

(i)  Curing time (100 g mass)  Greater than 40 minutes at 77 °F, with 4% cuemene hydroperoxide (ASTM D 2471)

(j)  Bond strength greater than 1,500 lb/in$^2$ (ASTM C 882)

(k)  No unreactive solvents or diluents shall be permitted in the monomer system.

(2) Initiator-promoter system. -

    (a) Initiator Cuemene hydro peroxide - 78 percent

    (b) Promoter Cobalt Napthenate - 6 percent

The initiator/promoter system shall be capable of providing a surface cure time of not less than 40 minutes nor more than 3 hours at the surface temperature of the concrete during application. The initiator/promoter system shall be such that the gel time may be adjusted to compensate for changes in temperature that may occur throughout the treatment application.

(3) Sand. - The sand shall be clean, dry, and free of organic materials, silt, and clay. Except as otherwise approved by the Contracting Officer, the sand shall conform to the following grading:

| U.S. standard sieve size | Percent passing |
|---|---|
| 4.75 mm (No. 4) | 100 |
| 2.36 mm (No. 8) | 90-10 |
| 850 μm (No. 20) | 5-10 |
| 300 μm (No. 50) | 0-10 |

This grading is intended to allow the use of commercially available silica sands of No. 8/20 or No. 10/20.

e. Safety. - All work shall be performed in accordance with the requirements of paragraph 1.6 and these specifications. The materials shall be stored, handled, and applied in accordance with the manufacturer's recommendations. Storage of all materials shall be in the original shipping containers and as specified in this paragraph.

<u>INITIATORS AND PROMOTERS SHALL BE STORED SEPARATELY SINCE COMBINATION CAN RESULT IN A VIOLENT REACTION OR EXPLOSION.</u>

Personnel exposed to monomer, initiator, or promoter, or their vapors shall use minimum protective equipment as follows: safety eye glasses, impervious gloves and aprons, and rubber boots as required. As determined by the Contracting Officer, personnel may be required to use full-face protective shields, self-contained respiratory equipment, or both. All personnel handling the monomer or catalysts shall be thoroughly trained in their safe use in accordance with the manufacturer's recommendations.

Unsafe handling practices will be sufficient cause to discontinue work until the hazardous procedures are corrected. The handling and use of the monomer and catalysts shall in all cases comply with the requirements of applicable Federal, State, and local safety requirements and ordinances.

The Contractor shall provide an eye wash and water washing facility for use in the event of accidental splashing of the monomer or catalysts on the workers. Eyewash facility shall be capable of providing a clean, room temperature flushing stream for a minimum of 15 minutes. All sources of sparks or flame must be removed from areas used for storage and handling. In these areas storage and mixing vessels shall be provided with electrical grounds to prevent static sparks.

Mixing and transfer equipment shall be explosion proof, and sufficient ventilation shall be provided to prevent the formation of explosive sealant/air mixtures. In accordance with applicable safety regulations, warning signs, such as "No Smoking" signs, shall be posted. In public areas, care shall be taken to eliminate sources of sparks or flame when the monomer system is present. Particular attention should be given to removal of welding operations, posting of "No Smoking" signs, and to traffic control to eliminate accidental fires from these sources. Visitors at the job site should be warned of the potential hazards and provided with applicable safety equipment. The Contractor shall also place an adequate number of 4-A:60B:C fire extinguishers on the job, so that no portion of the monomer system application is conducted farther than 100 feet from the nearest fire extinguisher.

f. Concrete preparation. - The concrete surfaces to be treated shall be clean, dry, and physically sound. All deteriorated concrete shall be removed in accordance with paragraph 2.1 to obtain a physically sound surface for treatment. The Contractor shall perform all work required to bring the surfaces to this condition.

Concrete surfaces shall then be prepared by power sweeping and by blowing with high-pressure air to remove all dirt and foreign material from the surface and from all cracks. Contaminants, such as asphalt and heavy oil and rubber stains, shall be removed, at the discretion of the Contracting Officer, by scraping and cleaning with solvents. Well-bonded surface contaminants and existing painted surfaces shall be removed by wet sandblasting. High-pressure water blasting will be permitted only if it can be demonstrated to the satisfaction of the Contracting Officer that the concrete surface and cracks can be completely dried prior to the application of the polymer treatment.

g. Application. - The methacrylic concrete sealing compound shall be mixed, applied to the prepared concrete surface, and cured in accordance with these specifications.

(1) Mixing of materials. - The monomer shall be mixed with initiator and promoter in the following proportions (proportions may be adjusted by the Contracting Officer to give a satisfactory pot life based on concrete surface temperatures and recommendations of the monomer system manufacturer):

|  | | Parts by weight |
|---|---|---|
| (a) | Substituted Methacrylate Monomer | 100 |
| (b) | Cuemene Hydroperoxide (78%) | 4.0 |
| (c) | Cobalt Napthenate (6%) | 2.0 |

<u>DO NOT MIX COBALT NAPTHENATE DIRECTLY WITH CUEMENE HYDROPEROXIDE AS THIS WILL PRODUCE AN EXTREMELY VIOLENT AND EXPLOSIVE CHEMICAL REACTION.</u>

The cobalt napthenate and cuemene hydroperoxide shall be mixed with the substituted methacrylate monomer in separate steps; i.e., first add and mix cobalt napthenate with the substituted methacrylate monomer, and then add and mix the cuemene hydroperoxide with the cobalt napthenate/ methacrylate monomer mixture.

The materials shall not be premixed. The monomer system shall be applied to the concrete surface within 5 minutes after mixing the cuemene hydroperoxide with the cobalt napthenate/substituted methacrylate monomer mixture.

For manual application, the quantity of monomer system mixed shall be limited to 5 gallons at a time. A significant increase in viscosity or change in gel time prior to application shall be cause for rejection.

Machine mixing and application of the methacrylic sealing compound may also be performed by using a two-part monomer system utilizing a promoted monomer for one part and an initiated monomer for the other part. Adequate mixing shall be done to achieve a uniform blend of the two parts.

The use of machine mixing and application requires approval by the Contracting Officer and treatment of a 10- by 50-foot test site to demonstrate that the equipment is working properly and capable of providing a uniform monomer mixture.

(2)  Application of sealing compound. - Prepared surfaces shall be protected from rain and moisture. Surfaces shall be treated with sealing compound within 24 hours after surface preparation is completed. The surface shall be allowed to dry thoroughly for a minimum of 48 hours before treatment. At the discretion of the Contracting Officer, the following test shall be made to determine if the concrete surface is sufficiently dry to proceed with the polymer treatment. A 2-foot-square piece of clear polyethylene sheeting shall be taped to the surface of the concrete and allowed to remain there for a minimum of 2 hours exposed to sunlight. Moisture condensation on the inside surface of the polyethylene sheeting shall be considered as evidence that the concrete surface is not sufficiently dry, and an additional period for drying will be required before proceeding with the polymer treatment.

Additional tests may be required, at the discretion of the Contracting Officer.

The monomer system shall be applied to the concrete surfaces during nighttime and early morning hours as directed by the Contracting Officer and with concrete temperatures between 45 and 85 °F. Monomer application will not be permitted in the direct rays of the sun as this may cause a premature curing of the system.

The concrete surfaces shall be treated with monomer at an application rate of 75 to 100 square feet per gallon. The concrete surfaces shall be flooded with the monomer mixture, allowing full penetration of the concrete and filling all cracks, and brushed with a stiff bristle broom. Puddles of excess monomer shall be removed by the Contractor.

The sealing compound may also be spray-applied by machine using a two-part mixing procedure described in subparagraph 3.12.g.(1). The pressure at the spray nozzle shall not be great enough to cause monomer mist to drift more than 2 feet beyond the nozzle. Compressed air shall not be used to produce the spray.

(3) Application of sand and curing. - Within 15 to 20 minutes after application of the methacrylic sealing compound, and before significant gelling has occurred, the entire treated area of concrete shall be covered by sand broadcast to achieve a uniform coverage of 0.25 to 0.50 pound per square yard. This sand shall be left on the concrete surface until the sealing compound has cured to a tack-free condition. Any excess sand, not bonded to the concrete, shall then be removed by the Contractor. Sand shall not be applied to vertical surfaces that have been treated with methacrylic sealing compound.

Treated areas shall be protected and not put back into service for 24 hours after treatment to allow the sealing compound to fully cure.

(4) Cleanup. - The Contractor shall keep the mixing equipment and tools clean during the course of the treatment, using a suitable solvent such as acetone or methyl ketone (both flammable), or 1,1,1,-trichloroethane (nonflammable). Soap and water are also satisfactory for cleanup of fresh monomer from the tools. The Contractor shall quickly clean up all spills by a method previously approved by the Contracting Officer.

After the repair work is completed, the residue in the open containers of cuemene hydroperoxide and cobalt napthenate shall be safely destroyed by using some of the excess monomer resin to wash out the catalyst containers and then allow it to cure before disposal or by other methods recommended by the manufacturers and approved by the Contracting Officer.

3.13    SURFACE IMPREGNATION

a. General. - These specifications present the requirements for impregnating concrete with a methyl methacrylate based monomer-catalyst system followed by in situ polymerization of the monomer by heat.

b. Submittals. -

(1) The Contractor shall provide the Contracting Officer with the manufacturer's certifications that the monomers and catalyst meet these specifications. Representative samples of the monomer system components, or of the combined monomer system if purchased premixed, shall be delivered to the Contracting Officer at least 30 days prior to use. At the Contracting Officer's option, these samples will be tested to determine specifications compliance.

(2) At least 30 days prior to beginning the concrete impregnation process, the Contractor shall deliver to the Contracting Officer a written report describing the Contractor's planned treatment procedures. Included in this report shall be a detailed description of the drying, impregnation, monomer mixing and storage, polymerization and quality control procedures, facilities, and equipment the Contractor intends to use to treat the concrete. The Contracting Officer will review this report and approve or disapprove the plan within 30 days of the date of receipt. In no event shall the Contractor proceed with the surface impregnation treatment until approval of the Contractor's procedures, materials, and equipment has been received.

(3) During the drying, cooling, impregnation, and polymerization cycles, the Contractor shall obtain and supply to the Contracting Officer concrete temperature data accurate to $\pm 5$ °F from at least nine points uniformly spaced on the surface of each treatment area and from at least one point 1 inch below the concrete surface at the approximate center of each treated area.

These data shall be in the form of a continuous record or periodic readings recorded at 1-hour intervals. The technique and equipment used to obtain the temperature data required in subparagraph 3.13.g.(1) shall be described in the written procedures report required above and subject to the Contracting Officer's approval.

(4) The Contractor shall maintain and supply to the Contracting Officer monomer and catalyst records listing the dates of manufacture, storage temperatures, date of use and application rates, and quantities as applied to the concrete.

c. Quality assurance. - Quality assurance shall be in accordance with paragraph 1.4 and these specifications. Representative material samples submitted by the Contractor, as specified in subparagraph 3.13.b.(1) will be tested, at the Contracting Officer's option.

d. Materials. -

M-47 (M0470000.896)
Page 59 of 73
8-1-96

(1) Monomer system. - The monomer system shall be composed of 95 percent by weight methyl methacrylate (MMA) and 5 percent by weight trimethlolpropane trimethacrylate (TMPTMA). A polymerization catalyst, 2,2-azobis-(2,4-dimethylvaleronitrile), shall be added to this monomer system at the rate of 1 part catalyst to 200 parts monomer by weight, or as directed by the Contracting Officer.

(a) MMA. - MMA shall meet the following requirements:

| | |
|---|---|
| Formula | $CH_2=C(CH_3)COOCH_3$ |
| Inhibitor | 25 parts per million hydroquinone (HQ) |
| Molecular weight | 100 |
| Assay (Gas Chromatography) % | 99.8 min |
| Density | 7.83 lb/gal (0.938 kg/L) |
| Boiling | 212 °F (100 °C) |
| Flash point (Tag, ASTM D 1310) | 55 °F (13 °C) |

(b) TMPTMA. - TMPTMA shall meet the following requirements:

| | |
|---|---|
| Formula | $(CH_2=CH_3 COOCH_2)_3CCH_2CH_3$ |
| Inhibitor | 100 parts per million hydroquinone (HQ) |
| Assay, % | 95.0 min |
| Density | 8.82 lb/gal (1.058 kg/L) |
| Flash point | Greater than 300 °F (149 °C) |

(c) The polymerization catalyst shall be 2,2-azobis-(2,4-dimethyl-valeronitrile). Empirical formula: $C_{14}H_{24}N_4$.

Monomer system components shall be used within 6 months after manufacture.

(2) Sand. - The impregnation sand shall be composed of clean, hard, dense, low-absorptive particles that will pass a 1.18-mm (No. 16) sieve, but with not more than 5 percent passing a 150 $\mu$m (No. 100) sieve.

e. Safety. - All work shall be performed in accordance with paragraph 1.6 and these specifications. Because of the hazards associated with improper use and handling of the monomer and catalyst, the following additional safety requirements shall be adhered to during the surface impregnation process.

Personnel working with the monomers or catalyst shall be provided with and use safety eyeglasses or goggles, impervious gloves, aprons, and boots. Normally, in an outdoor monomer application, respiratory equipment will not be necessary. In storage and mixing operations, however, accidental spills or equipment failures may result in hazardous vapor concentrations requiring self-contained respiratory equipment for personnel protection. The Contractor shall provide a field eye wash and water washing facility for use in the event of an accidental splash of monomer on the workers. Eye wash facility shall be capable of providing a clean, room temperature flushing stream for a minimum of 15 minutes.

All sources of sparks or flame shall be removed from areas for monomer storage and handling. In these areas monomer storage and mixing vessels shall be provided with electrical grounds to prevent static sparks. Mixing and transfer equipment and motors shall be explosion proof, and sufficient ventilation shall be provided to prevent the formation of explosive monomer vapor-air mixtures. In accordance with applicable safety regulations, warning signs such as "No Smoking" regulations shall be posted. At the construction site, care shall be taken to eliminate sources of sparks or flame when monomer is present. Particular attention shall be given to removal of welding operations, posting of "No Smoking" signs, and to traffic control to eliminate accidental fires from these sources. Visitors at the job site shall be warned of the potential hazards and provided with applicable safety equipment.

Catalyzed monomer not used within 4 hours of catalyst addition shall be stored in an explosion proof storage facility at a maximum storage temperature of 0 °F until it can be used or destroyed as approved by the Contracting Officer. Storage of catalyzed monomer for periods longer than 2 days will not be permitted.

Monomer and catalyst storage and handling. - The monomers, MMA and TMPTMA, or the premixed monomer system shall be stored in their original shipping containers or in other clean containers as approved by the Contracting Officer. Maximum monomer storage temperature shall not exceed 90 °F. The storage area shall be selected to provide protection from direct sunlight, fire hazard, and oxidizing chemicals. Sufficient ventilation shall be maintained in the storage area to prevent the hazardous buildup of monomer vapor concentrations in the storage air space. The polymerization catalyst shall be stored in accordance with the manufacturer's recommendations, but in no event shall the catalyst storage temperature be allowed to exceed 35 °F. Personnel exposed to monomer or monomer vapor shall use minimum protective equipment as follows: safety eyeglasses, impervious gloves and aprons, and rubber boots as required. As determined by the Contracting

Officer, personnel may be required to use full face protective shields, self-contained respiratory equipment, or both. All personnel handling the monomers or catalyst shall be thoroughly trained in their safe use in accordance with manufacturer's recommendations.

Unsafe handling practices will be sufficient cause to discontinue work until the hazardous procedures are corrected. The handling and use of monomer shall in all cases comply with the requirements of applicable Federal, State, and local safety requirements and ordinances.

During the polymerization cycle, the heating enclosure shall be provided with a means of positive ventilation to prevent hazardous concentrations of monomer vapor within the enclosure. Open flame heat sources will not be approved for use during polymerization.

The monomer mixing area shall be free of sources of ignition and shall be well ventilated. Spilled monomer shall be contained with absorptive materials such as vermiculite or dry sawdust and removed with non-sparking equipment.

f. Concrete preparation. - Deteriorated concrete shall be removed from the surface by wet sandblasting or other suitable means. Concrete containing surface contaminants such as oil, paint, or protective coatings shall be cleaned by wet sandblasting or by other approved means to remove these materials. After the removal of these materials, the concrete surface shall be swept and air-blown to remove sand, leaves, trash, gravel, or other miscellaneous loose materials to the satisfaction of the Contracting Officer.

The Contractor shall install a temporary dike along the high side of the area to be impregnated to divert possible rainwater around the area to be impregnated. The Contractor shall also install a temporary dike along the low side of the area to be impregnated to act as a barrier to prevent monomer from accidentally escaping due to an accidental spill or excess application.

g. Application of the surface impregnation process. -

(1) Drying. - After the concrete surface area to be treated has been cleaned in accordance with subparagraph 3.13.f., it shall be uniformly covered with a 1/4- to 1/2-inch thick layer of sand meeting the requirements of subparagraph 3.13.d.(2) and dried to permit polymer penetration. The equipment used to accomplish drying shall consist of a weatherproof enclosure with either an electric infrared, or hot-air heat system, or other technique as approved by the Contracting Officer. Exposed flame infrared heat systems, if selected for drying, shall not be used for polymerization.

Drying shall be accomplished by raising the concrete surface temperature at a rate not exceeding 100 °F per hour to between 250 and 275 °F, and maintaining that surface temperature range for 8 hours. If a higher maximum temperature is desired, approval by the Contracting Officer shall be obtained. During the drying period, sufficient airflow shall be maintained over the concrete surface to ventilate the water vapor removed from the concrete and to provide uniform concrete surface temperature. During the drying cycle, the Contractor shall obtain continuous or periodic surface temperature measurements, as specified in subparagraph 3.13.b.(3), at a sufficient number of locations over the heated concrete surface [normally one location per 100 square feet] to ensure temperature uniformity. The maximum temperature variation over the heated concrete surface shall not exceed ±20 °F of mean concrete surface temperature at the time measurements are taken.

(2) Cooling. - After the concrete has been dried, it shall be cooled prior to monomer application. The cooling rate for concrete surface temperature shall not exceed 100 °F per hour. Cooling shall continue until the maximum temperature at a depth of 1 inch below the surface of the concrete is 100 °F or less.

During the cooling and impregnation cycles, the dried concrete shall be protected to prevent moisture from reentering the concrete. It may be necessary, if determined by the Contracting Officer, to repeat the drying and cooling cycles prior to monomer application should moisture reenter the concrete.

(3) Monomer mixing. - The monomer MMA and TMPTMA may be premixed in the specified ratio and stored prior to use. Storage of premixed monomer shall be as required in subparagraph 3.13.e. All monomer mixing and transfer equipment shall be as required in subparagraph 3.13.e. All monomer mixing and transfer equipment shall be of explosion proof design and shall be provided with electrical ground cables. Monomer transfer shall be from bottom to bottom of the vessels or through dip pipes in the vessels to prevent the buildup of static charge during transfer. Pipe fittings, valves, pump impellers, or other equipment which will come into contact with monomer shall not be made of copper or brass or certain plastics attacked by the monomer.

(4) Catalyst-monomer mixing. - The polymerization catalyst shall be mixed with the monomer system immediately prior to use. Monomer system temperature at the time of catalyst addition shall not exceed 90 °F. Mixing shall be accomplished with explosion proof equipment in electrically grounded containers in a well-ventilated area.

(5) Impregnation. - Following the drying and cooling cycles, the sand on the concrete surface to be impregnated shall be uniformly leveled if necessary.

The temperature on the surface of the concrete shall not exceed 100 °F at the time of monomer application nor at any time during the impregnation cycle.

Monomer application shall be made at a rate sufficient to uniformly saturate the sand layer to a slight excess without applying so much monomer that it would drain away from the impregnated area. The monomer application rate should be 0.8 pound of monomer per square foot of concrete surface. This is approximately 0.9 gallon per square yard of surface area. However, sand layer thickness, sand particle size, and slope may necessitate application rate adjustment to achieve the described saturation. Following application, the monomer shall be allowed to soak into the concrete for approximately 6 hours. If at any time during the soak cycle the sand should become dry, additional monomer shall be applied as directed by the Contracting Officer.

In order to protect the monomer-saturated sand from the polymerizing effects of direct and indirect solar radiation, monomer application and subsequent soaking shall occur during the time period sunset to sunrise unless the Contractor provides shielding, as approved by the Contracting Officer, to prevent solar radiation from reaching the area being impregnated. Immediately following monomer application, a continuous mylar membrane a minimum of 6 mils thick shall be placed over the monomer-saturated surface to reduce monomer evaporation. This membrane shall remain in place, except for the short periods of monomer application or surface inspection, throughout the impregnation cycle and until the polymerization cycle is complete.

(6) Polymerization. - Polymerization of the monomer impregnated into the concrete shall be accomplished by uniformly heating the treated concrete to a surface temperature of at least 165 °F and not exceeding 185 °F and maintaining it for a minimum of 5 hours. The rate of temperature increase and allowable surface temperature variation shall be as required in subparagraph 3.13.g.(1).

The equipment and procedures used to accomplish heating of the concrete for polymerization shall be of the type described in subparagraph 3.13.g.(1) as approved by the Contracting Officer. See subparagraph 3.13.b.

(7) Cleanup. - Following completion of the surface impregnation treatment process, the Contractor shall remove the sand from the concrete surface and dispose of the sand at the site as directed by the Contracting Officer.

3.14 SILICA FUME CONCRETE

a. General. - Silica fume concrete shall be used to repair concrete damaged by abrasion-erosion action. Silica fume concrete may also be used in the infrequent occasions where a high strength (compressive strength in excess of 10,000 lbs per square inch) repair concrete is required. Silica fume concrete shall be used on areas of damaged concrete greater than 1 square foot having a depth greater than 6 inches or a depth extending 1 inch below or behind the backside of reinforcement. If the depth of repair is at least 2 inches but less than 6 inches, epoxy bonding agent shall be used in accordance with the provisions of paragraph 3.8, to bond fresh silica fume concrete to concrete being repaired. Silica fume concrete shall not be used for repairs that are less than 2 inches in depth.

b. Submittals. - The Contractor shall submit certification of compliance for materials in accordance with subparagraph d. below.

c. Quality assurance. - Quality assurance shall be in accordance with paragraph 1.4.

d. Materials. - All concrete materials shall be obtained from previously tested and approved sources. Materials will be accepted on certificate of compliance with the following ASTM Standards:

(1) Portland cement. - Portland cement shall meet the requirements of ASTM C 150 for type I, II, or V cement. The specific cement type shall be as directed by the Contracting Officer and determined by the environment in which the repair is conducted.

(2) Silica fume. - The silica fume mineral admixture shall be obtained as a byproduct from the manufacture of solely silicon metal in electric blast furnaces. The condensed silica fume shall be processed and sized to a fineness of approximately 200,000 $cm^2$ per gm (20,000 $m^2$/kg) at a porosity (t) of 0.500 when tested in accordance with ASTM C 204 and have an amorphous silica ($SiO_2$) content of not less than 85 percent of the total fume. When tested in accordance with ASTM C 311, the silica fume shall have a moisture content of less than 3 percent and a loss on ignition of not greater than 6 percent. A manufacturer's certificate of compliance with these requirements and applicable provisions of ASTM C 618 is required. The silica fume shall be supplied, proportioned and combined with other admixtures, as necessary, from a supplier regularly engaged in the sale of this combination product as a concrete admixture. This combination admixture shall be batched with the concrete in either of two forms or types.

(a) The wet type shall consist of water slurry containing approximately 45 percent silica fume solids with a water-reducing admixture meeting all requirements specified.

(b) The dry form shall be a densified powder blended with a dry water reducing admixture. Both types shall be compatible with a water reducing admixture that could be added at the concrete plant or at the placement site.

(3) Admixtures. - The Contractor shall furnish air-entraining and chemical admixtures for use in concrete.

(a)  Air-entraining admixture shall be used in all silica fume concrete and shall conform to ASTM C 260.

(b)  Chemical admixtures. - The Contractor may use type A, D, F, or G chemical admixtures.  If used, they shall conform to ASTM C 494.

(c)  Use of other admixtures must be approved by the Contracting Officer.

(4)  Water. - The water used in making and curing silica fume concrete shall be free from objectionable quantities of silt, organic matter, salts, and other impurities.

(5)  Aggregate. - The term "sand" is used to designate aggregate in which the maximum size particle will pass a 4.75-mm (No. 4) sieve.  The term "coarse aggregate" is used to designate all aggregate which can be retained on a 4.75-mm (No. 4) sieve.  Sand and coarse aggregate meeting the requirements of ASTM C 33 shall be used in all concrete.

(6)  Curing compound. - Wax-base (type I) and water-emulsified resin-base (type II) curing compounds shall conform to the requirements of Reclamation's "Specifications for Concrete Curing Compound" (M-30) dated October 1, 1980.

(7)  Evaporation retarder. - Monomolecular membrane evaporation retardant formulated for use with silica fume concrete requirements shall be equal to "Confilm", manufactured by Master Builders, Lee at Mayfield, Cleveland, OH 44118.

e.  Safety. - All work shall be performed in accordance with paragraph 1.6.

f.  Concrete preparation. - After damaged or unacceptable concrete has been removed as specified in section 2. the surface on which the silica fume concrete will be placed shall be prepared.  An acceptable surface shall have the appearance of freshly broken, properly cured concrete.  The surface shall be free of any deleterious materials such as free moisture, ice, petroleum products, mud, dust, carbonation, and rust.  The perimeters of the repair shall be saw cut to a minimum depth of 1 inch.

The clean surface is not ready to receive repair silica fume concrete until it has been brought to a saturated, surface-dry condition.  This condition is attained by saturating the surface to a depth that no concrete mixture water may be absorbed from the fresh concrete.  Then, just prior to placing concrete against the surface, all free moisture (moisture capable of reflecting light) shall be removed from the prepared surface.

g.  Application. - Silica fume concrete shall be composed of cement, silica fume, coarse aggregate, sand, water, and approved admixtures, all well mixed and brought to the proper consistency.  Silica fume concrete mixtures shall be proportioned in accordance with Reclamation's "Concrete Manual",

Eighth Edition, revised, chapter III, except that silica fume shall be added to the mixture at a ratio of 7 to 12 percent by mass of the portland cement as directed by the Contracting Officer. The water-cementitious ratio of the concrete (exclusive of water absorbed by the aggregates) shall not exceed 0.35 by weight. Slump of the silica fume concrete, when placed, shall not exceed 3 inches for concrete in slabs that are horizontal or nearly horizontal and 4 inches for all other concrete. Silica fume concrete with less slump should be used when it is practicable to do so. The concrete ingredients shall be thoroughly mixed in a batch mixer. The concrete, as discharged from the mixer, shall be uniform in composition and consistency from batch to batch.

(1) Forms. - Forms shall be used for silica fume concrete whenever necessary to confine the concrete and shape it to the required lines. The forms shall be clean and free from encrustations of mortar, grout, or other foreign material. Before silica fume concrete is placed, the surfaces of the forms shall be coated with a form oil that will effectively prevent sticking and will not soften or stain the concrete surfaces or cause the surfaces to become chalky or dust producing.

(2) Placing. - Placing of silica fume concrete shall be performed only in the presence of an authorized representative of the Contracting Officer. Placement shall not begin until all preparations are complete and the authorized representative of the Contracting Officer has approved the preparations. Silica fume concrete shall not be placed in standing or running water unless, as determined by the Contracting Officer, the structure under repair cannot be economically dewatered. If underwater silica fume concrete placement is required, special placing procedures shall be required. A suggested guide is ACI 394R.

When appropriate, silica fume concrete shall be placed in layers not greater than 20 inches thick. Each layer, regardless of the thickness, shall be adequately consolidated using immersion-type vibrators or form vibrators when approved. Adequate consolidation of silica fume concrete is obtained when all undesirable air voids, including the air voids trapped against forms and construction joints, have been removed from the concrete.

(3) Finishing. - The class of finish required shall be a finish closely resembling the finish of the surrounding concrete. Silica fume concrete does not normally develop bleed water and special finishing procedures may thus be required. The ambient temperature of surfaces being finished shall be not less than 50° F. Immediately following placement of silica fume concrete to finished grade, the surface shall be screeded to bring the surface to finished level with no coarse aggregate visible. No cement or mortar shall be added to the finishing operation.

A monomolecular membrane evaporation retarder shall be applied to the surfaces of the silica fume concrete, in accordance with manufacturer's recommendations, immediately after the screening operation.

Floating, if necessary to achieve the specified finish, shall be performed immediately following the application of evaporation retarder.

h.  Curing and protection. - Proper curing of silica fume concrete is essential if bond failure and shrinkage cracking are to be eliminated. Silica fume concrete repairs shall be cured, preferably by water curing, or alternately, by application of a uniform and continuous membrane of wax-base (type I) or water-emulsified resin-base (type II) curing compound meeting the requirements of subparagraph 3.6.d.(6). and as approved by the Contracting Officer.  If the use of curing compound is approved, daily inspection by the Contractor shall be performed to ensure the maintenance of a continuous, water-retaining film over the repaired area.  The water-retaining film shall be maintained for 28 days after the concrete has been placed.

Silica fume concrete surfaces to which curing compound has been applied shall be adequately protected during the entire curing period from pedestrian and vehicular traffic and from any other possible damage to the continuity of the curing compound membrane.  Areas where curing compound is damaged by subsequent construction operation within the curing period shall be resprayed.

Water curing shall commence immediately after the concrete has attained sufficient set to prevent detrimental effects to the concrete surface.  The concrete surface shall be kept continuously wet for a minimum of 14 days. Whenever possible, silica fume concrete shall be water cured by complete and continuous inundation for a minimum period of 14 days.

The Contractor shall protect all silica fume concrete against damage until acceptance by the Government.  Whenever freezing temperatures are imminent, the Contractor shall maintain the newly placed repair concrete at a temperature of not less than 50 °F for 72 hours.  Water-cured silica fume concrete shall be protected from freezing for the duration of the curing cycle and an additional 72 hours after the water is removed.

3.15  ALKYL-ALKOXY SILOXANE SEALING COMPOUND

a.  General. - A concrete sealing compound is defined as a liquid that is applied to the surface of hardened concrete to prevent or decrease the penetration of liquid or gaseous media, such as water, aggressive solutions, and carbon dioxide, during the service exposure, preferably after initial drying to facilitate its absorption into voids and cracks. An alkyl-alkoxy siloxane sealing compound, herein after referred to as siloxane, shall be applied to concrete surfaces when it is desired that application of a sealing compound cause no change in the appearance of the sealed surfaces.

The sealing of concrete surfaces with siloxane shall be in accordance with these specifications.

Siloxane sealing compound shall not be used to seal concrete subject to frequent or permanent immersion in water; nor shall it be used on concrete surfaces exposed to high abrasion forces.

b.  Submittals. - The Contractor shall, before starting work, shall submit to the Contracting Officer manufactures data and certification that the concrete cleaner and siloxane sealing compound furnished by the Contractor meets the requirements of this specifications.  The chemical constituents shall correspond to the requirement of subparagraph 3.15.d.

A MSDS shall be furnished to the Contracting Officer prior to shipment of material with information pertaining to the safe practices for storage, handling and disposal of the materials and their explosive and flammable characteristics, health hazards, and the manufacturer's recommended fire fighting techniques.  The MSDS shall be posted at all storage areas and at the job site.

The Contractor shall furnish, for approval, a detailed written description of the methods the Contractor plans to use to apply the siloxane and for clean up of all spills and residue in open containers.

c.  Quality assurance. - Quality assurance shall be in accordance with paragraph 1.4.

d.  Materials. - The siloxane sealing compound shall be a clear, ready to use sealer based on oligomeric alkyl-alkoxy siloxane containing not less than 20 percent active siloxane solids by mass.)  The compound when properly applied to concrete shall conform to the following performance standard:

| | | |
|---|---|---|
| Chloride screening | 91% (minimum) | (NCHRP 224, Series I) |
| Reduction of Chlorine Penetration vs untreated concrete | 92.4(minimum) | (NCHRP 224, Series IV) |
| Moisture Vapor Transmission | 97.5%(minimum) | (ASTM E-96) |
| Water Repellency Rating | 92%(minimum) | (ASTM C-140, ASTM C-67) |
| Water Absorption | 1.40%(maximum) | (ASTM C-67, ASTM C-140) |
| Scaling Resistance to Deicers | Excellent | (ASTM C-672) |
| | | |
| Resistance to Chlorine Penetration | 0.07 lbs/cu.yd. (maximum) | (AASHTO T-259/260) |
| Surface Friction Reduction | 0 | (ASTM E-303) |
| Penetration (1 application) | 1/8" - 1/4" | |

The compound shall have a high flash solvent carrier with a strong chloride screen and shall exhibit alkaline stability and form a chemical bond with the treated concrete.

The concrete sealing compound shall be Consolideck SX as manufactured by ProSoCo Inc., P.O. Box 1578, Kansas City, KS 66117, (913) 281-2700, or approved equal.

e.  Safety. - The Contractor shall take the necessary precautions to avoid wind drift onto auto and pedestrian traffic.  The materials shall be stored, handled, and applied in accordance with the manufacturer's or supplier's recommendations.  Storage of all materials may be in the original shipping containers.  The material should be stored in sealed containers and kept away from extreme heat.  The sealant contains blended solvents and should be handled accordingly.  Do not use near fire or extreme heat and provide good ventilation to avoid buildup of solvent fumes.  Personnel applying the concrete sealant shall wear NIOSH/MSHA approved respirators, goggles, rubber gloves, and plastic or rubber suits to avoid splash to skin and eyes.  Clothing that becomes contaminated with the concrete sealant shall be changed as quickly as possible.

Unsafe handling practices will be sufficient cause to discontinue work until the hazardous procedures are corrected.  The handling and use of the concrete sealant shall in all cases comply with the requirements of applicable federal, state, and local safety requirements and ordinances.

f.  Concrete preparation. - The concrete surfaces to be treated shall be clean and physically sound.  Unsound or deteriorated surfaces shall be removed in accordance with the requirements of section 2.  All needed repair work shall be adequately cured prior to application of the siloxane.

Concrete surfaces shall be prepared by power sweeping and blowing with high pressure air to remove all dirt and foreign material from the surface and from all cracks.  Contaminants, such as asphalt and heavy oil and rubber stains, shall be removed at the discretion of the Contracting Officer by scraping and cleaning with solvents.  Well bonded surface contaminants and existing painted surfaces shall be removed by high pressure water blasting or wet sand blasting.  The concrete surface and all cracks shall be completely dried prior to the application of the siloxane.

g.  Application. - If the concrete sealant is a product other than Consolodeck SX, a test application of the concrete sealant shall be made to an area selected by the Contracting Officer using the same equipment and procedures proposed for the project.  The test procedure is to insure compatibility of the product, to determine the waterproofing results and to check for surface discoloration from the procedure.  The Contractor shall not proceed with the remainder of the work until the Contracting Officer approves the results of the test application.  If the results of the test application are deemed unsatisfactory by the Contracting Officer, the Contractor shall modify his sealant materials, procedure, and/or equipment as directed by the Contracting Officer and the test application shall be repeated.

The siloxane sealant shall not be applied at surface and air temperatures below 40 degrees F, or above 100 degrees F. Surfaces shall be treated within 24 hours after the surface preparation is completed. The prepared surfaces shall be maintained in a dry condition and protected to prevent contamination prior to the siloxane application. If the prepared surface becomes contaminated, it shall be recleaned in accordance with paragraph 3.15.f. The Contracting Officer shall determine if the surface is dry enough to receive sealant. If the Contracting Officer determines that the surface is too damp for application of the sealant, application of the sealant shall not commence without approval of the Contracting Officer.

The sealant shall be applied with low pressure (20) psi airless spray equipment fitted with solvent resistant hoses and gaskets. Heavily saturated brush or roller may be used in isolated incidents if the Contracting Officer determines that brush or roller application is the most effective means.

Adjoining glass, metal and painted surfaces shall be protected from overspray and splash of the siloxane sealant. Any accidental or unintentional overspray or splash on adjoining glass, metal or painted surfaces shall be removed using mineral spirits before the solution has dried on the surface.

When applying to exteriors of occupied areas, all exterior air conditioning and ventilation vents shall be covered during application and air handling equipment shall be turned off during application to avoid solvent odors within the occupied areas.

The concrete surfaces shall be given a full and complete application of the siloxane sealant at the following application rate of 80-120 sq.ft./gal.

(1) Horizontal surfaces. - When applying the siloxane to flat horizontal concrete surfaces, the siloxane shall be applied in two "wet-on-wet" coats. Flood the surface and broom or squeegee the material around for even distribution. Allow the surface to absorb the siloxane and follow immediately with a second application before the surface dries. Puddles of excess siloxane sealant shall be broomed out thoroughly until they completely penetrate into the surface.

(2) Vertical surfaces. - When applying the siloxane to vertical surfaces, the siloxane shall be applied in two "wet-on-wet" applications. Apply the siloxane in a flooding application, from the bottom up with sufficient material applied to produce a 6" to 8" rundown below the contact point of the spray pattern with the concrete surface.

Allow the first application to penetrate the surface (approximately three to five minutes) and reapply in the same saturating manner. If the siloxane sealant is applied to surfaces of extremely dense, mirror finish concrete the Contracting Officer may direct that the siloxane be applied in one saturating application to prevent surface darkening.

h. Curing. - The treated areas shall be protected from rain and foot traffic for six hours after application. Vehicular traffic will not be allowed on the treated area until after 24 hours after the application of the siloxane.

i. Measurement for Payment. -

(1) Measurement for payment of surface preparation of the concrete surfaces will be made of the actual surface area prepared. Payment for surface preparation will be made at the unit price per square foot bid therefor in the schedule, which unit price shall include all costs of preparing the concrete surfaces for the siloxane as specified in paragraph 3.15.f.

(2) Payment for seal coating of the concrete surfaces will utilize the same area as measured for the surface preparation. Payment for seal coating of concrete surfaces will be made at the unit price per square foot bid therefor in the schedule, which unit price shall include all costs for storing and handling materials, of applying the siloxane system, of cleanup, and of providing all the necessary safety equipment, and any other work required under these specifications to properly complete the job.

Notes

www.ingramcontent.com/pod-product-compliance
Lightning Source LLC
Chambersburg PA
CBHW061325190326
41458CB00011B/3902